CHANGING
THE WORLDS

CHANGING THE WORLDS

The FOR-PROFIT Plan to Mine Asteroids and
Terraform Two Planets in One Human Lifetime

JOHN AMABILE

ARCHWAY
PUBLISHING

Archway Publishing books may be ordered through booksellers or by contacting:

Archway Publishing
1663 Liberty Drive
Bloomington, IN 47403
www.archwaypublishing.com
1 (888) 242-5904

ISBN: 978-1-4808-0717-4 (sc)
ISBN: 978-1-4808-0718-1 (e)

Library of Congress Control Number: 2014938573

Print information available on the last page.

Archway Publishing rev. date: 10/5/2016

Contents

Introduction

Before I delve into the nuts and bolts of my plan, I would like my readers to open their minds. Only a tiny part of this plan is solely mine; much of it was pieced together by men and women before me. Having said that, this book will be like nothing you have ever read.

Imagine the first man to behold a piece of driftwood. Eventually someone took the bold step of altering a piece of wood and created the first canoe. Perhaps he did this for territory, to impress a girl, for property. Maybe it was the simple joy of using his mind to alter his surroundings. There is very little chance that it was because a tyrant told him to. Even Christopher Columbus approached the Spanish monarchy, not the other way around. But imagine what a disaster it would have been for mankind if a witch doctor had met the man on the beach and told him, and everyone that followed him, that he could go out to sea to look at fish and birds but that it was strictly forbidden (by whatever unjust authority) to derive any material benefit from his enterprise. Imagine pasteurization, antiseptics, the Bessemer process, steam power, aviation, metal, agriculture, and fire never opened to economic activity but forever made the "collective property of mankind."

It is impossible to get the consent of 7 billion people. To give something to mankind means depriving all individuals, corporations and national governments from using it. I believe that mankind is a good thing. I disagree with *The Hitchhiker's Guide to the Galaxy* when it said that the creation of the universe was a bad idea. Since mankind is a good thing, its continued existence is also. It is possible that we will manage our resources well enough to forever live on Earth. Since we live in hundreds of different states, and we each have our own individual mind, we cannot blink in unison. Who thinks 7 billion people can unanimously plan a sustainable move to outer space? If you put someone in a room where the only door had to be opened by a billion people at the same time, surely we would call such a place prison.

In 2011, I flew completely around the entire earth. I was shocked at how simple a thing this was: except for a few space probes, all of humanity's existence can be traversed in about two days and for a few thousand dollars.

Isn't it amazing to think that in four hundred years a civilization as wide as the solar system will look back on the twenty-first century as one with no vision of the future, stifling economic growth, and silly conflicts over pocket change, compared to the resources that will then be available to mankind?

Indeed, this is the view that we hold of the world during the Dark Ages, and this disparity was created because a handful of individuals created a state that is responsible for almost all political and technological achievements of the modern world.

When America gained its independence in 1789, mankind did not have electricity (Edison), airplanes (Wright brothers), usable oil (Rockefeller), moving assembly lines (Ford), modern banks, (JP Morgan), cheap steal (Carnegie), atomic power (US Department of Defense), computers, Internet, and iPhones. In America, government was very limited. Business was free to solve the problems of the market, we flew over some problems, beamed X-rays through others, we looked solutions up on google, or asked our iphones. All the governments of earth have copied the solutions the free market has developed, not the other way around.

War and conflict come from one government drawing a line around a resource, continent, or ethnicity and threatening to bomb, shoot, or bludgeon anyone who passes their line. In 1967, all the governments of Earth came together and drew such a line in the sky.

The Outer Space Treaty, conveyed by the media as banning all weapons in space, made it impossible for any individual or government to acquire real estate in space. Henceforth space was the "collective possession of mankind." Imagine where our species would be if cars, books, or food were banned from private ownership, and you will have answered the question of why people have failed to colonize space: our grandparents made it illegal. What good results have ever come from transferring property from private to public hands?

Johann Gutenberg took out a loan from a bank to create the first printing press. Henry Ford was so successful creating the first mass produced car that virtually every manufactured item on earth is made using his method. Without the idea of private property, economic growth is impossible. This is why the real estate market outside of Earth has grown at 0 percent for the last forty years.

Under the Outer Space Treaty of 1967, platforms launched into orbit are the property of the launching entity, so the tiny belt of territory above Earth has seen billions of dollars of investments and trillions of dollars of income (cell phones, GPS, television, and Internet-bearing satellites). But the rest of the universe is closed to free enterprise and anything that depends on free enterprise to prosper (*like homo sapiens sapiens*).

That might sound very strange, particularly since most space-faring governments have spent hundreds of billions of dollars in space and have made no progress toward space colonization since 1972. I'm not accusing anyone of making a profit from colonizing space. I'm accusing *EVERYONE* of not making a profit from colonizing space. So, to skip the political stuff, the only way mankind can successfully colonize the solar system is if we directly profit from it.

What if North America was uninhabited in the seventeenth century and England banned any merchants, farmers, or fishermen from crossing the Atlantic? Only a few aristocrats, funded by the crown were allowed to cross and at huge expense. Perhaps after six successful expeditions (and one where they ran out of drinking water and had to turn back), they would have beaten the French in an expensive game of Capture the Flag, collected some New World rocks, posed for a painting (no businessmen having yet created cameras), sailed back to England, and never returned. The English could brag about this accomplishment for decades, but that is small compensation for abandoning a new world. If there were no expectations of profit to make investors bold, then sixteenth-century England would have been the height of freedom, wealth, and technology for humankind. Where would we be?

I propose something that is philosophically, militarily, and economically (I'm ignoring politically) correct: the complete privatization of space and the formation of a Space Corporation and an Extraplanetary Exchange for the sale of property off of Earth. The Space Corporation will have a few initial tasks in order to open the solar system up for settlement, but as much of this process will be privatized as possible and as soon as possible. Space Corporation can sell, lease, or rent this property to other entities through the Space Exchange, which will be the simplest way of generating short-term money.

When England (before it was the United Kingdom) set out to establish colonies around the world, she did this in the form of trade companies. Ohio, Virginia, India, Pakistan, Bangladesh, South Africa, Singapore, even the Dutch in

Indonesia did not originally organize these colonies directly from their government. The businesses that organized them picked economically viable places to settle; they factored in logistics, politics, trade routes, and resources. I am not condoning the colonization of the world by Europeans, but the state-centered colonies organized by Portugal, Spain, Italy, and France have been rife with civil war and stifled economic growth while the business-centered colonies established by Britain and Holland have fared much better. South Africa is the richest and most stable country in Africa. Singapore is one of the richest countries per capita in the world. America is the richest and most stable country that has ever existed. The Virginia and Ohio Companies organized investors, funded immigration, selected wise locations on ports and rivers, and dealt with most of their shareholders through free exchange and voluntary choice.

In the former Spanish colony of Bolivia, there have been 192 coups since they gained independence in 1825. That wasn't 192 years ago. Haiti has had forty-eight presidents since their independence from France, and only two of them finished their terms. The poverty of these societies is directly tied to their political instability and the use of force in political decision making, a thing they preserved from their colonial forebears.

In contrast, business is a much more stable, peaceful entity than government. There is no need to hold a rebellion or a protest if someone doesn't like his or her local grocery store; you just go to the next one. If Space Corporation is managed poorly or unethically, people are free to not buy its products.

There is all the difference in the solar system between consenting citizens managing their communities and an administrative area of coercion. America and Britain have had very small armies and police forces relative to their populations for the last several centuries. They are also two of the wealthiest nations on Earth.

Aside from a few properties set aside for initial terraforming and colonization, any citizen, business, or democratic state can acquire property in the solar system. The Space Exchange will simply keep accountability of who owns what, and ensure that the purchasing entity is, or is from a democratic state. Most of the company's revenue will come from short-term mining, which will make it profitable within ten years of founding the company.

Space Corporation will be a profitable asteroid-mining business at first, but the technology and operations for this mining effort will seamlessly transition to terraforming Mars and Venus. Terraforming will also be done for profit, but these profits will require more patience. Mining can return money in large quantities within eight years, while Mars and Venus will not be terraformed for about forty years. Since Mars and Venus will eventually be sold after they are developed, this is simply planetary real estate development. Terraforming is something to increase the value of property (like painting a house or hiring a landscaper).

Much of the effort of terraforming Mars and Venus can be done at little cost, since the terraforming operations will be conducted in coordination with mining operations. We will have most of the technology built and operational, already mining asteroids. As I will outline in the rest of my plan, a solar-system-wide mining survey, smelting and refining capability and a large amount of excess mining material will directly facilitate terraforming operations.

What financial cost that terraforming will have, independent of the space-mining effort, will be sustained by a 10 percent funding rate put on Space Corporation, in lieu of paying taxes. This will also be profitably encouraged by the Space Exchange. As each step in terraforming is achieved, the expectation of selling Venus and Mars will encourage investors to get in as early as they can. Other than rescinding the Space Treaty and passing the Space Corporation Charter, the governments involved will support this plan by giving obsolete military technology to Space Corporation in exchange for dividend-yielding stock.

If someone in England four hundred years ago said he wanted to build a country that had ninety times the population of England (at the time), with ten thousand towns and three of the largest cities on Earth, his countrymen would have laughed at him. That undertaking would have been impossible or cost quadrillions of British pounds. This has been built nonetheless, and not by a king at bayonet point. A tiny regulatory presence of the state, twenty million small businesses and thousands of big businesses accomplished something all governments forever would not have been able to.

Private businesses branching off into subsidiary corporations in space will be encouraged, as long as they are from democracies and have certain low standards of compatibility in their equipment. Great care will be taken to ensure this doesn't become a monopolizing force. The docking bays will have to be interoperable but will not have to be made by the same company, just as most soda cans come in one standard size for convenience but are made by different entities. Docking and communications equipment will be the main things regulated. Comparable safety standards for civilian aviation on Earth will apply in space.

As Europeans moved west across America from the Atlantic coast, there was a constant cycle of buying cheap, undeveloped land, developing it, and then selling the land at a high price and moving farther west. This same phenomenon will propel the exploration and colonization of space.

The ore in orbit will be sold to customers on Earth, and then it will enter the normal domain of taxation. So when Space Corporation makes one billion USD from the first mining expedition, some of the money will cover the operational costs, some will go to the shareholders in the form of a stock dividend, and 10 percent will go to the terraforming budget.

Once terraforming is completed and the planets are auctioned off piece by piece, Space Corporation will make a profit from this as well. The global rare earth metals market will move through Space Corporation hands, and the only sectors for the company to grow will be extraplanetary real estate development. The 10 percent stipulation will keep Space Corporation focused on the long-term goal. The corporation may even want to spend more than 10 percent, but that is the minimum. Solid dividend performance will raise the stock price, and inefficiency and accidents will lower it.

I have not concerned myself with how my plan may sound to the majority of people. I have only been concerned with the physics, economics, philosophy, and history of it. I'm proud to report that the math worked out, so here is my plan.

CHAPTER 1
General Overview

My plan for colonizing space begins with the American government—and any other democracies that want in—rescinding the Space Treaty of 1967, which bans the privatization (and hence profitization) of space. This same treaty also bans weapons in space (mostly it's the weapons ban that's touted to the public.) Forcing the entire universe to be unprofitable is an absurd idea, no matter who supports it. I wouldn't mind a different treaty preserving the weapons ban. Only democratic states would be allowed in space, and America would be business partners if not military allies with all of them. If we signed an Ocean Ban Treaty and a Land Ban Treaty, everyone on Earth would starve to death.

We will found the first multiplanetary corporation, which I am naming Space Corporation. Space Corporation will own anything past the Earth's Atmosphere except the sun and human-owned things already in space. States and corporations will be offered stock shares for existing debris in space. This is all voluntary—if the Russians want one million paint chips circling in low Earth orbit at twenty thousand miles per hour forever, they can keep them. If they want to sell the mass (and the legal liability) to Space Corporation for a few shares of stock, it's their call.

I will use the metric system of measurement throughout this book. Ironically, the US military has been using the metric system as part of the NATO military alliance for decades. Marksmanship examinations and maps in the US military are done in the metric system. It is also frankly a more scientific system for technical writing. Other than a few useful developments in computers and robots, the newest technology in this book is forty years old. Capitalism, the core part of my plan, is thousands of years old, and I'm sure that is the part that will sound the most novel to many of my readers.

As an American, I have written this work from a largely American viewpoint. As someone whose grandparents volunteered to be Americans, I want to emphasize that Space Corporation will be as American as electricity, computers, and airplanes. If humankind makes soft drinks and cars in free-market, multinational corporations, there is no reason we should not develop real estate and mine metal in free-market, multiplanetary corporations.

There are many private space companies today (Planetary Resources, SpaceX, Virgin Galactic, and Deep Space Industries, to name a few). Space Corporation will function as a real estate development and mining corporation and will be free to contract out to these other space businesses or to take part in joint ventures. The corporation could take the form of ten different companies with reciprocal agreements (one makes engines, another landing gear, etc.). We will also have a space exchange set up to sell shares of any property offworld. Companies can branch off or merge as much as they wish. An important note here: When I use the word 'we' in this book I don't mean mankind, or the US government. I mean people who consent to participate in colonizing space. I believe it is unethical to force human beings into a pronoun.

When they look at our solar system, many people are saddened at how bleak and harmful our family of planets is to life. But if you show a real estate developer a vacant lot, he isn't despondent; he's building condos in the back of his head! I argue (and will in the rest of this book) that terraforming, the process of making these worlds more earthlike, is far from being wild fantasy. It is not even that hard. The key question is, simply, will we allow men and women to profit from interplanetary real estate development?

Only citizens of states that sign the Space Corporation Charter will be allowed to purchase from or sell to Space Corporation. A state must have had orderly elections for ten years to qualify. Humanity's movement to space will take a long time, but it would be a wonderful thing for a democratic Cuba, a (real) Republic of China, or a (real) Republic of Iran to come with us to the stars.

FIRST STEPS

The corporation will trade shares of stock to any democratic government on Earth in exchange for Cold War–era ballistic missiles (with warheads removed), nuclear fuel, some existing space agency facilities, and eventually atomic devices. The atomic devices (reengineered atomic weapons) will be used for breaking up asteroids very far from Earth, and they will be escorted by members of the contributing nation's military all the way to their destination.

Intercontinental ballistic missiles (ICBMs) are important because these systems are free-delivery vehicles to low Earth orbit. Hundreds of these weapons are slowly decaying in the weapon stockpiles of Russia and several NATO powers. Thousands of these weapons were built during the cold war. They are simply disassembled or destroyed in their silos. Each and every one of them is a free ride to earth's orbit.

We will eventually use atomic devices in space, and these can also be exchanged to the corporation for stock. Members of whichever military contributes the weapons will escort the machines to their destination so there will be no security risk. Actual detonation codes can be kept on Earth and beamed at the last second, which is already done for ballistic missile submarines.

Russia has more atomic weapons than all of the other nations of Earth combined.[1] Russia is already selling the uranium nuclear fuel in their warheads to US nuclear power companies, so this is not a far-fetched idea. If no foreign country wants to participate, the United States can easily contribute sixteen aging ICBMs. As a former officer in the US military who worked with missiles and rockets, I can assure everyone that we have a huge supply of these systems. As they age, they are disassembled or fired without warheads into oceans and deserts for target practice. America could contribute the oldest sixteen of these rockets, and it would have no significant impact on our military preparedness vis-à-vis the Russians (since this is only 3 percent of our strategic weapons).

Russia will benefit more from this than America will, so I am assuming Russia will contribute eight of their R36 rockets to match our eight UGM-133 Trident rockets. Russian rockets are much larger than the standard US rocket. Both rockets will carry a space probe built around a 5MW nuclear thermal rocket

---

1   The lowest estimates I've seen of stored nuclear weapons are eight thousand for Russia and seven thousand for the United States. Everyone else combined has about 1,000. http://www.businessinsider.com/nine-nations-have-nukes-heres-how-many-each-country-has-2014-6

and several other sensor pods. This basic probe I am naming a Kata (from *Katakopos*, Greek for scout).

The R36s, with three times the lifting capacity, will carry a Kata and a special probe for releasing sensor balloons to float around the Martian atmosphere. The smaller American ICBMs will carry one Kata and additional smaller probes for the gas giants. The purpose of the Katas is to collect data for terraforming and space mining at the same time.

This phase could be easily done for $5 billion USD in cash as long as the government technology was traded for stock shares and businessmen and scientists were working to keep costs down. NASA successfully tested a "space-worthy" nuclear thermal rocket design in 1973, but the program was scrapped, and the facility has been vacant ever since. We could acquire this equipment and expertise in exchange for stock from the US government. As a by-product of the nuclear arms race, there is a large amount of nuclear fuel owned by various states on Earth, far more than we will use for the civilian power industry in the foreseeable future.[2] Nuclear thermal rockets (NTRs) will be the workhorses of this entire effort outside of the atmosphere. NTRs run on fuel we paid for decades ago, and they need very little of it. We can move all of the initial small pieces of equipment into Earth orbit using ICBMs. While I only mention sixteen ICBMs in my plan, we could certainly benefit from more of these systems, particularly if the United States and Russia were to coordinate taking a small fraction of their missiles offline and exchanging them to Space Corporation. The only reasonable military target for the Intercontinental ballistic missiles of the

---

2   There are thirteen thousand plutonium bomb pits in storage in the United States. I used the number 11,000 throughout this book to account for age and decay. Russia probably has even more, since they have more atomic weapons and lower quality plutonium than the US. http://www.lasg.org/campaigns/PUPitProd.htm

United States and Russia are each other. I think we'll find a sky full of gold to be a stronger incentive for reducing nuclear arms than any hitherto proposed.

Once the mining information comes back from the Katas and other probes, we will know the locations for large concentrations of easily recoverable precious metals and terraforming resources. This will make the stock price of Space Corporation increase, and more funding will be available for the next phase. I also believe that the enthusiasm for space across the world will generate a huge amount of investment, even without the expectation of profit (profits will bring in non-space enthusiasts).

It is very rare for NASA to use even two space probes with similar designs, which drives up development costs exponentially. As a government organization, no matter how noble the intentions of NASA, they have no profit motive to keep them focused, to ensure high quality, and to keep down operating costs. This has led to the public perception that all activity in space is expensive.

I propose to use sixteen probes with an identical design and integrated subcomponents across related reconnaissance technology as well. All of the robots used for moving ore later will be identical to the Katas, except they will have large mechanical arms added to them and fewer sensory modules.

Four years after Space Corporation is founded, we will launch the first Katas. The Katas will have several targets each to scan en route to the Kuiper Belt. We will launch all of the probes over a two-year window. Eight of the Katas will fly near Mars and release a pod that will enter the Martian atmosphere. These probes will have balloons and will float around Mars for two years, scanning the surface for minerals, CO_2 ice, and water ice. Four other Katas will drop gas giant pods into the orbits of the four gas giants. Each pod will scan tens of thousands of asteroids and dozens of moons around each planet. Finally, four of the rockets that carry Katas will bring simple utility robots, called proto-ballistae, into low Earth orbit to start gathering up space debris.

The Katas will deploy a small radio tag to any Kuiper Belt objects that are the right size and composition for terraforming activity. The gas giant pods will help retransmit radio signals from the Katas back to Earth.

Once the information from the Katas has been collected, we will start mining small asteroids as they come near Earth. As our technology improves, we will be able to wade out a little farther into the ocean of space. Eventually we'll fly to larger asteroids, remove valuable chunks, and fly the pieces back to Earth orbit to be smelted. Early on we will concentrate mining on one asteroid that is 10 or more kilometers across and make this our shipyard.

After the smelter, we will launch large unmanned craft called ballistae (singular *ballista*, after Roman stone-throwing machines). These will be almost identical to the first sixteen Katas, except they will not have the various sensory attachments and instead will have a set of large mechanical arms. These arms will have heating elements on the ends so they can melt into and then out of cold space objects (like ice asteroids). For convenience, I will use the near-Earth asteroid 433 Eros. After careful examination, we might find a better one, but this is a good starting point.

The smelter and ballistae will already be in orbit, gathering space debris. They can even be sent ahead of the human crew to save time. The only new piece we will need is an asteroid torch for melting out pieces of asteroids. We will need a human command-and-control ship, which will work both in space and, with a few extra pieces, on the surface of Mars or Venus.

We will keep costs down on our mining expedition by bringing a small ice asteroid to Earth orbit from the main asteroid belt. Our first mining operations will be near Earth, controlled by radio signals from the ground. We will build up space infrastructure from debris and mined asteroids. In particular, we will make a propellant station and an orbital farm (a green station). Water from the main belt can be electrolyzed to produce liquid hydrogen propellant. Hydroponics can be used on our green station to grow crops.

With food and propellant waiting for us in orbit, the cost of a three- or six-month mining expedition will largely go down. We can generate oxygen from water ice and all drinking and hygiene water; even the food the astronauts eat can consist largely of water by mass.

By melting down asteroids in orbit, we will separate the ore into different elements. Any element that is worth more than US$10,000 per kilogram I consider marketable metal. These metals will be smelted into small shapes, called ore bundles, and sold to customers on Earth. They will be delivered to our customers from space by GPS guided parachutes. The machine used to accurately and safely bring an ore bundle to earth is called an ore bundle parachute assembly (OPA). As space commerce expands we can shift from 10,000 USD per kg down to 5,000 USD per kg, down to 1,000 etc. The main limiting factor will be the parachutes bringing the ore to the surface. In the long term, anything that costs more than US$100 per kilogram could be profitably imported from space.

Most of an asteroid, by mass, will be unmarketable material. We have to separate and melt this material to get to the valuable parts. Since we're planning for the long term, we will build the nonmarketable metal into hull plates, support beams, radiation shielding, and whatever else our intrepid heart's desire. At first this will be simple things but eventually entire fleets and cities will be made from these by-products.

The Habitation module (Hab) will be large enough to comfortably fit a four-person crew.[3] Virtually all of the different systems the crew control will be unmanned robots and the HAB will have control interfaces for monitoring and directing these robots. The HABs will never come back to Earth. The HABs will either remain in space on asteroid mining expeditions or they will have extra parts added and land on the surface of Mars or Venus.

Movements to and from the surface of all planets will take place in single-stage-to-orbit (SSTO) craft. We will have slightly different versions for each planet. Once the crews are in space, they will be pushed back to Earth by a ballista. Except for the crews and billions of dollars in precious metals, what goes up will not come down.

Meanwhile, after our sixteen Katas have arrived in the Kuiper Belt, we will pick four methane/nitrogen Kuiper Belt objects (KBOs) with a mass of roughly ten billion metric tons each. At year ten, the sixteen Katas will begin to push these objects toward Uranus. The probes will push the objects for ten years into a gravity assist. A gravity assist basically uses the gravity of a large object to change the velocity of a smaller object – like bouncing a ball off of a speeding train. During the gravity assist we can radically change the velocity of these objects (velocity is both direction and speed). The KBO asteroids will coast through space for twenty years after the gravity assist and then impact Mars during the south polar summer in year forty.

My inspiration for this idea comes from an article written by Dr. Robert Zubrin.[4] I have added it to my for-profit plan

---

3   Robert Zubrin, *The Case for Mars* (Touchstone: Simon & Schuster, 1996), 7.

4   See http://www.users.globalnet.co.uk/~mfogg/zubrin.htm.

since it can be performed concurrently with profitable mining operations and technology. We will direct these impacts to the areas we've found to have the highest concentrations of frozen CO_2 and water ice from our Mars balloon probes.

The reason we will plan this for the Martian south polar summer is because the Martian atmosphere is ten times its normal thickness in the Southern Summer. The south has more ice and frozen CO_2 than the north. When the southern ice melts, it thickens the atmosphere. This is the best time to heat the crust to Earth level. Bringing thousands of asteroids down will flash evaporate both the material imported from space and the frozen material currently on the Martian surface.

We will also coordinate several ice-meteor or comet impacts from the main asteroid belt at the same time as the nitrogen objects arrive. Mars will eventually need a great deal of water to be earthlike, and importing the water in such a dramatic fashion will aid in warming the planet. Since we know all the materials that Mars needs to be earthlike, we will simply move these materials from different parts of the solar system and import them to Mars.

We will be careful about the order we are adding new elements. When water ice is added to Mars on top of the CO_2, the effect is solid carbon, water, and oxygen (this is called the Bosch reaction). Methane-asteroids can be brought from the Kuiper Belt as needed, but the ball will be rolling.

After profitable mining is underway, we will have enough money to adjust the orbits of objects already in the Martian L4 and L5 Lagrange points. We will add huge reflective surfaces to these objects to increase sunlight reaching Mars. LaGrange points are areas where the gravitational attraction between two objects of significant mass is neutralized, creating a stable "third orbit" for objects of negligible mass. The L4 and L5 points are not along the line of sight for Mars, so we can reflect sunlight from these positions and direct it onto the day side of Mars. There are a total of three asteroids one kilometer or more in diameter currently in these two positions, as well as several smaller objects that can be used for this purpose.

We will have to build the support structure and the reflective surface and attach them to the Martian LaGrange objects. We will also have to neutralize their rotations relative to the sun and Mars. A 5% increase in sunlight on Mars will cause Mars to start slowly warming. By standardizing all of our equipment, we will already have these support beams and aluminum reflectors while mining ice for Venus as well. We will use large pieces of aluminum foil to cover ice asteroids going to Venus. Later this aluminum foil can be moved to the Martian Lagrange objects and used to reflect additional sunlight to Mars.

We will also build a Helmholtz coil at the Martian L1 point. The L1 point is a permanent position locked in line of sight between the sun and Mars. Helmholtz coils are metal coils attached to electric generators that produce electromagnetic fields. These fields can be used to neutralize electromagnetic waves, such as the electromagnetic radiation coming from the sun. This is the largest source of radiation in our solar system, and blocking it will make Mars much more livable. All of the sunlight will be allowed through (the coil is hollow in the middle, resembling a bicycle tire). On the outside of the coil we will place solar panels to provide power, but they will not be along the line of sight from the sun to Mars, so no light will be blocked.

By placing a large Helmholtz coil at the L1 point, we will place Mars forever in the radiation shadow of the coil. The coil would have to be of enormous size (twenty-four thousand miles in diameter). There is probably more than twenty-four thousand miles of Internet cable in a small American town. This structure will be made from unmarketable material at virtually no cost. We can terraform *while* mining.

Since the impact event will not occur until the fortieth year of Space Corporation, we will have plenty of time to build these objects. The important thing is to plan each of these separate activities with the larger goal in mind.

In the fifth year of Space Corporation, the first ballista will arrive in the main asteroid belt, and it will divert a large ice asteroid (about ten thousand tons) to high Earth orbit.

Once the ballista arrives in Earth orbit, we will basically have a small lake's worth of water permanently at the top of the gravity well. We will sell some of this water to other governments and space businesses (it costs US$10,000 per kilogram to ship water from the surface to Earth orbit). Some of this water will go to Earth Green Station 1. Green Station 1 will be made from debris in Earth orbit. As we expand into space, we will build more and larger green stations. A key part of our colonization of the solar system will be constructing green stations wherever we plan to go. Eventually we will make much larger green stations (called mega-green stations) for seeding planets with life and growing food for small cities in space.

The first green stations are designed only to feed a four-person crew. They will be very small and cheap and will not be permanently manned. With missions occurring every few months, each crew will board one, harvest the crops, plant a new crop, and then leave.

If we have water in space, we can also use electrolysis to generate liquid hydrogen in space. This is the propellant that is used by nuclear thermal rockets. The energy will come from salvaged solar panels already in space, or from nuclear reactors carried into orbit. As more and more nuclear fuel is expended in space, we will plug the diminished low-energy fuel cells into these systems to add to the steady rate of propellant production. Most of the mass needed to travel to an asteroid is liquid hydrogen propellant, and all that we need to acquire this in space is liberty, frugality and imagination.

Access to food, fuel, water, and air in space will create a strategic shift in the logistics and financing involved in colonizing the solar system. When an American gets on a plane and flies to Europe to study for a year, he doesn't have to bring all of his food, water, air, and fuel with him. Only a tiny number of billionaires, or yet more extravagant government employees, could afford to make such a journey.

All of these initial steps can be put in place without any manned missions. NASA frequently operated the robotic arm on the space shuttle from the ground while the crew in space slept. Even with all of these cost-saving measures, some emergency supplies will be kept in space, and most complex technological items will also be flown up from Earth. The goal of Space Corporation will be to export (to Earth) more dollars than we import.

Having astronauts and scientists farm may seem more heretical to some people than earning a profit or using atomic devices without killing people. The Romans planted agricultural colonies throughout their republic, and the British came to America looking mainly for farmland. Our species has a symbiotic relationship with plants; unless we bring them with us, the Milky Way galaxy will remain an expensive photo opportunity.

ATOMIC MINING

While ballistae moving small asteroids to Earth will provide us with hydrogen fuel, food, and a solution for instant coffee in Earth orbit, we will need a different method for ice on a planetary scale. Terraforming Mars and Venus will require a huge amount of ice, about 1 percent the mass of Earth's moon. There are many sources of water ice in the solar system, but I have selected the dwarf planet Ceres as the best and nearest source of water outside of the gravity well.

For this plan, I propose the bold step of cutting out large pieces of ice from Ceres and removing them with fusion devices. A fusion device is basically a reengineered hydrogen bomb. We will excavate these large pieces of ice at the proper trajectory to send them to Mars or Venus. We will use large space manufactured ships (ballista-2s) to make minor course adjustments after the initial excavations. Any man-made asteroids or asteroids whose courses are altered by humanity will be called a "pili" (from Roman throwing spear) to distinguish them from naturally formed asteroids. Mining will use pili (less than one kilometer in diameter) while terraforming will use megapili (larger than one kilometer in diameter).

Professor Paul Birch calculated that 4E+19 kilograms of water ice would be enough to reduce the CO_2 level on Venus and bring the temperature down to Earth norm.[5] I plan to build a Helmholtz coil that is 50 percent opaque at the Venusian L1 point. This will have the effect of blocking all of the harmful radiation Venus receives from the sun, as well as 50 percent of the sunlight (matching it to Earth's light level).

This object will resemble a large ceiling fan, with a coil around the outside. The fan blades will be made of solar panels, and they will block 50 percent of the radiation and 50 percent of the sunlight. The open sections between the blades will let sunlight through but not radiation. Since the blades are made from solar panels, they will have a strong electrical current that will be used to make an electromagnetic field, repelling radiation.

---

5   http://www.orionsarm.com/fm_store/TerraformingVenus
     Quickly.pdf

My projections in this book are based on moving 3.6E+20 kilograms of ice to Venus (this much ice has 4E+19 kilograms of hydrogen). This will be enough water to remove almost all of the CO_2 from Venus. We will do all of the chemical reactions between imported ice and the Venusian atmosphere first, which require heat. After we get the composition of gas where we want it, we will block half of the sunlight to lower the temperature so it's comparable with Earth.

Now, moving 3.6E+20 kilograms of ice is a big task. Fortunately, our ancestors left us about eleven thousand very powerful tools. To be perfectly clear, I am *not* speaking of our functioning nuclear arsenal. The American armed forces have 2,100 atomic weapons that can be launched or dropped on our enemies with little notice. Similarly, the Russian armed forces have 1,800 atomic weapons in this same category. These eleven thousand tools I am referring to make up the US stockpile of plutonium bomb cores, which were never actually put into bombs and are being guarded in very expensive warehouses.

I am also not counting three thousand partially disassembled US systems and eight thousand partially disassembled Russian systems. Eventually the absurdity of maintaining such a large number of nonfunctioning weapons will become apparent, but now I am only referring to the separate pieces of plutonium. Since a nuclear war would be over in a matter of hours, this doesn't even make sense from a purely military perspective. It is really caused by the fact that these weapons were expensive to make and more expensive to throw away safely. My readers must also appreciate how much less a government is accountable to their shareholders than a business.

On Earth, it is very expensive to guard and maintain atomic weapons. There is considerable fear that some of these Russian weapons may be sold on the black market, and there's always the small risk of accidents or theft. A black market is by definition unquantifiable, but providing a free-market alternative would at least raise the black market price of an atomic weapon. Since the weapons will stay in the physical possession of that nation's armed forces, (even in space), there will be no legitimate security risk. A small number of these weapons have already been decommissioned and used for civilian power (the civilian Megatons to Megawatts Program). Nonetheless, this is better for Russian uranium weapons; we would have to spend even more money to convert plutonium bomb pits back into uranium (making them much weaker).

Even a five-kiloton bomb (which is the butter knife of our atomic arsenal) has enough energy to vaporize fifteen cubic kilometers of ice. In space, whenever something is melted, it is also propelled, because the vacuum of space will naturally carry material in the direction of the vacuum. While we have never detonated atomic bombs for mining ice, the United States and the former USSR conducted many belowground atomic tests here on Earth, and we have an enormous amount of data about blast depth, blast radii, and effects on different soil compositions.

There will be a small amount of radioactive material mixed with the ice by using this method, but it will not be enough to affect our colonization of Venus or Mars. We have detonated many atomic bombs on Earth's surface, and while I am not condoning this practice, I am also alive to tell about it. My estimate is that eleven thousand fusion devices made from these plutonium bomb cores will be enough (one at a time) to move the ice that Venus needs. Mars may need as few as 20 devices.

We will have 5,500 Ice Megapili cut out of Ceres or other asteroids and comets. The objects will be guided through space for several years, and then another fusion device will be used to break them into smaller pieces that will burn up in the Venusian atmosphere.

Mars can be terraformed in the same general way but on a smaller scale. Venus needs 3.6E+20 kilograms of water, Mars requires 4E+18 kg of nitrogen and 1E+18 kg of water. I'm planning for another 1E+19 kilograms of water to Mars from small debris while terraforming Venus. Suffice to say Mars will need about %1 the mass that Venus will need. There are millions of asteroids of various sizes in the main belt, Kuiper Belt, and in the orbits of our various planets. There is already a lot of water and CO_2 on Mars—we just need to add more and heat up what is there already.

Many of these atomic weapons are aging—barring species wide genocide, they'll go to waste! In the next twenty years, they will become garbage too dangerous to throw away. Figuring out the right spot to detonate these weapons and getting them to that spot will be very complex. But the physics itself was worked out in the early 1950s.

Of course, this part of the plan will sound controversial. Atomic weapons are supposed to be used at great expense, on Earth and against humans. Not profitably, millions of miles away from Earth, and on inanimate objects. But since I've been outside of the box for some time, I'll just keep going. This method of mining ice will make it possible to alter the climates of Venus and Mars in one human lifetime. It will also provide an honorable, safe, and profitable way for nations to discard some of their atomic weapons. Even if Atomic ice mining failed, we'd be better off for having tried.

To prevent national security concerns, the nations that contribute nuclear devices would train members of their own military, who would be attached to space corporation for any mission involving nuclear devices. These devices will be partially disassembled during their launch in order to prevent detonation in the event of an accident. The components will only be rendered operable in space. Detonation codes will be withheld by that government until the day they are used (which is the case with atomic weapons on Earth). There have been many ships with nuclear weapons that sunk, or planes with nuclear weapons that crashed. None of these weapons detonated, they didn't even leak.

While the nuclear thermal rocket is proven, good technology, we will also invest money in developing Variable Specific Impulse Magneto-Plasma Rockets (VASIMIR) and Nuclear Salt Water Thermal Rockets (NSWTR). Flying to Ceres by NTR takes 6 months, with better propulsion we can cut this in half, maybe more. Any manned voyage near Saturn will be very dependent on fast travel. This development will also reduce travel time in the inner solar system. For a business time is money. A six week trip to Mars is much cheaper than a three month or six month trip to Mars. Manned missions to the outer solar system are only conceivable with faster propulsion than NTRs. Each successful step in this process will lead to an increase in the stock price for the Space Corporation. Instead of committees of thousands of politicians, many of whom are rivals or even enemies, we will have small numbers of businessmen, engineers, and accountants dealing with each other for mutual advantage. When we're done, we will have doubled the habitable area for humanity in the universe, placed hundreds of nuclear-powered vessels in the heavens, changed humanity's economy and energy consumption, and placed the democratic nations of Earth so far above the dictatorships that in fifty years our grandchildren will think the word *tyrant* was always a synonym for *clown*.

The rare minerals that we will procure while we do this will drastically change humanity's economy. Computers, engine components, nuclear fuel—technology of all kinds will be cheaper as their component elements become more abundant. New York's skyscrapers could not have been built without Carnegie's cheap steel. Manufacturing will flourish as precious metals become common. Meanwhile, the most dangerous and pollution-causing economic activity can be shifted offworld.

CHAPTER 2

Equipment Development

Much of this plan is built on existing technology but nonexistent equipment. All of these pieces would be mass-produced, so they would all have standard parts and be interoperable. I have assigned numerical designations to these systems for the ease of technical writing. All items will use the same power plant (the ESR-1, see the next page). If power requirements are too great, multiple ESR-1s will be used. One of the reasons NASA is astronomically inefficient (pun intended) is they have few mass-produced items and instead custom design most of their equipment. Probes and scientific inquiry on far off celestial objects requires customization. There is no profit motive behind NASA (except for the publicity of novel achievements). Building 200 probes to go to Titan is much more cost effective than building 200 unique probes for dealing with the gravity, atmosphere, distances and orbits of the 200 moons in our solar system.

NASA may transition away from exploration, but there will be plenty of space for NASA. The increase in human activity in space could have NASA inspect space vehicles, atomic mining, and pili trajectory. Even rescue operations would be best suited for a government organization (Space Coast Guard? I'm terrible at names). If two companies have worker-robots that crash into each other in orbit around 21 Lutetia it would be nice to have NASA representatives around to write the crash report and order one of them to pay a fine. Of course depending on who signs the space corporation charter and specific legal and financial relationships we may merge national space agencies. We could also have NASA responsible for rescues and legal disputes near Martian Space with ESA performing the same function near Venusian Space. If Russia takes too long to sign their space agency will be rescuing robots near Pluto.

NASA launch capability can be purchased in exchange for stock. Even with making bulky items in space there will be a huge demand for expertise, launch hardware, and launch facilities. It will be several years until we are able to make complicated components in space. The SLS rockets can carry hundreds of high mass objects. We can have the occasionally repurposed ICBM carry smaller items. The SLS rockets can move much of the hardware we need to colonize the solar system, instead of letting two astronauts do a space suit high-five on a barren world.

We do have a profit motive. Our mining survey can be conducted by a small number of identical probes. Development costs can be ten times the production costs for a single unit, hence the huge cost of our government's space monopoly. We will research and study, develop good equipment, and then mass-produce to bring down the cost. Businesses have been doing this for three hundred years; it was at the heart of the Industrial Revolution and the rise of the American and British economies.

An Englishman invented the assembly line, and an American invented the moving assembly line. The idea caught on, and now virtually all property that any person owns on this planet was made by this method. It can work just as well in space. Costs for spare parts, fuel pellets, and development must all be kept down if moving to outer space is to be profitable. We will go to space on Profit, not rockets. The rockets will merely be a business expense.

All of these pieces will be designed with the same connections so they can be modular and ad hoc combinations of different systems will be very easy to create. A habitation module will be able to add a Gravimetric Radar Scanner (GRS) or an Extra Surface Reactor (ESR), based on the mission. This will be accomplished by using identical docking and power pieces on all equipment. Updates to these systems will be planned across Space Corporation. We will make them interoperable with the old systems while we gradually introduce new ones. So an ESR-2 with four times the power output will still plug into a GRS-1, and so on. We will not have a system that when upgraded no longer functions with any integral components. (I've heard this referred to as an "up-downgrade.") Military technology is designed to have as many interoperable components as possible, even as new components and systems are introduced.

Component list for Space Corporation:

Nuclear thermal rocket (NTRO-1): This is a five-thousand megawatt engine (successfully tested by NASA in 1973 and deemed space worthy). We will obtain all designs, patents, and machinery for these from NASA in exchange for stock.

Extra Surface Reactor (ESR-1): This nuclear reactor is designed for use in space. It can power a five-thousand megawatt NTRO, plus 10 percent for additional attachments.

Maneuver Thrusters 1 (MT-1): This will be the primary propulsion for the proto-ballistae, gas giant pod, and Single Stage to Orbit (SSTO) vehicle. An MT-1 is the smallest nuclear thermal rocket we can make. These will be used extensively on the proto-ballistae, and they will be incorporated into other designs as a secondary system to the main NTROs. The MT-1 will have adjustable nozzles (similar to a jet with a vertical takeoff and landing on Earth). Abrupt directional changes and slight course variation will be easy to do with an MT-1. Secondary power units will be designed to provide energy for an MT-1, plus 50 percent.

Secondary Power Unit (SPU-1): These will be used for primary power on proto-ballistae, balloon pods, and gas giant pods. They will also provide backup power on HABs, Katas, smelters, asteroid torches, and ballistae. The SPU is a radioisotope electric generator (RTG, or RTIEG). It produces very little power but has low weight and low complexity and can last for years.

Standard Fuel Pellet (SFP-1): All equipment will use the same sized nuclear fuel pellets. The SPU will use a small number of SFPs, but the ESRs will use many. We will also standardize these pellets for use in robots and reactors on planetary surfaces.

Gravimeter/Radar Scanner (GRS-1): This is a device used for scanning objects in space and on the Martian surface. This same technology was used by the US military to scan Afghanistan for mineral deposits from low flying aircraft. It's also frequently used by oil companies. A gravimeter will detect the mass of an object, the radar will detect shape. When mass and shape are figured out we get density. The dense things on the periodic table are also the expensive things.

Human habitation module (HAB-1): This will be a two story-tall structure designed for a four-person crew to live in while controlling robotic equipment. These will be mass-produced and carried up on heavy-lift rockets. This idea is based on Robert Zubrin's HAB from *The Case for Mars*. The HAB by itself will allow humans to live and control different machines for mining in space. When we want to land the HABs on Mars or Venus, we will add more attachments to them for that purpose.

Habitation Landing Array (HLA-1 and HLA-2): This is a parachute assembly, based on the balloon pod assembly but including landing gear and a small rocket. When an HLA is attached to a HAB, the HAB will be able to land on a planet (though not take off). We will have two versions of HLAs: HLA-1 for Mars and HLA-2 for Venus.

Communication Satellite (CMS-1 and CMS-2): These will be small satellites that will establish an MPS (Mars positioning system) and eventually VPS (Venusian positioning system) constellation over our two real estate projects. They will be nearly identical to each other to save development costs, except that they will have solar panels that can be adjusted based on the amount of sunlight at their location. CMS-1 will be used for Mars, and the CMS-2 used for Venus. All satellites will be placed 30 degrees above and below the equator. We'll plan for all of the impactors to come in along the equatorial plane.

Balloon Scout Assembly (BSA-1): This will be a heat shield with a very large Mylar balloon inflated by pressurized gas. After this assembly gets to a certain thickness of atmosphere, the balloon inflates, the heat shield and gas canisters fall off, and the balloon begins to float around Mars.

Optic Spectrum Device (OSD-1): This is a small set of high-quality digital and infrared cameras, specifically designed for use in space under various light and radiation conditions.

Celestial Propellant Package (CPP-1): This is equipment to attach to a frozen, potentially gaseous object, which scoops up easily excitable matter to be heated as a propellant for use in an NTRO. The NTRO has a built-in propellant cell for short trips in the vicinity of Earth, but for long-duration voyages, it will need in situ propellant and will have a CPP-1 attachment.

Radio Tag System (RTS-1): This is a dispenser with a hundred lightweight radio transmitters, each of which has a tiny heating agent. Basically this is a fuel pellet that has a small radio attached to it. The radioactive element will be the smallest increment of fuel pellet (one SFP.) Once it attaches to an asteroid, the RTIEG will produce electricity and send out a small radio signal with a simple message 'beacon 01234, high uranium content.' All our communication satellites and gas giant pods will be designed to listen for and track these signals. All Space Corporation vehicles will carry one RTS to be activated as an emergency beacon.

Celestial Smelter (CS-1): This is a smelter designed specifically for use in space. Proto-ballista will move matter into the smelter, and the vacuum of space will be used to siphon matter out. It will be powered by an ESR-1, have a set of MA-1s for moving small pieces of ore, and have an MT-1 for propulsion. The CS-1 can also be used for thermolysis and electrolysis of water.

Asteroid Torch (AT-1): This is a torch designed for use on rocky asteroids. It will have different power settings for different asteroid compositions. This torch will be powered by an ESR-1, and it will have large mechanical arms (MA-2s) for moving the objects it cuts free and an MT-1 for getting into position. The main purpose will be to slice/melt off manageable pieces of ore that fit inside the smelter. One design for the Asteroid torch is using a heated tungsten rod. It will liquefy low melting point metal, sucking it into space and moving it out of the way.

Katas (KT-1): These are the scout ships we will send out to scan the entire solar system for minable asteroids and objects useful for terraforming. Components include the NTRO-1, GRS-1, RTS-1, SPU-1, OSD-1, MT-1, and CPP-1. While they will be virtually identical to ballistae, Katas will not have mechanical arms. These will be the first sixteen craft launched and will do all of the initial scouting in addition to pushing the KBO impactors to Mars.

Balloon Pod: This is a BSA-1 connected to a GRS-1 and an SPU-1. This will enter the Martian atmosphere, deploy a balloon, detach the gas canister and the reentry shield, and float around Mars for two years, conducting thorough scans of the Martian surface for minerals and frozen gas deposits.

Gas Giant Pod (GGP): This assembly will be released into the orbit of each of the gas giants. It will consist of an MT-1, SPU-1, GSR-1, OSD-1, and RTS-1. The GGP will float around a gas giant's orbit, cataloging objects too small to see from Earth. There are so many objects orbiting the gas giants that we cannot observe or scan them all while speeding by in a Kata at eighty thousand kilometers an hour. The GGP will have MTs to move around. Radio tags will be placed on particularly promising objects. The GGPs will also help to retransmit signals back to the interior of the solar system from the various Katas and the radio tags the Katas place in the Kuiper Belt.

Mining in Space

Mechanical Arms 1 (MA-1): These are the arms used on a proto-ballista. These are for grabbing small pieces of space debris. Initially used to clean up debris in Earth orbit, they will also work for the first small-scale mining operations.

Proto-ballistae (PB-1): These are small spacecraft used for gathering debris in low Earth orbit. The proto series will be for getting the software and the mechanical dexterity figured out. Proto-ballistae will have MT-1s, MA-1s, and an SPU. These will be used for moving small items around in zero gravity.

Ballista (BB-1): This is the bigger brother of the proto-ballista. Each will be propelled by an NTRO. Each will have an MT-1 (for close maneuvering), an MA-2, a ESR-1, and a CPP-1. As a backup system, it will carry an SPU.

Mechanical Arms 2 (MA-2): These will be based on the MA-1 design, but they will be much larger and have the ability to heat up. This ability will help the arms move ice pieces. The "hands" will have the ability to heat up so that it can melt into a piece of ice. Afterward, it can be reheated to release the piece of ice.

Ore-bundle parachute assembly (OPA-1, OPA-2, and OPA-3): This is a parachute attached to an altimeter, with an Earth

GPS and small cameras with transmission capability that deploy along the nylon straps attached to the parachute. These will be used to guide the ore bundles in from Earth's atmospheric entry onto a conveyor belt at a smelter on earth. We will have different versions for different planets (OPA-1 Earth, OPA-2 Mars, and OPA-3 Venus).

Earth Single State to Orbit vehicle (ES-1): This will look like a small space shuttle. On the ground on Earth, 90 percent of the weight of this craft will be rocket propellant. It will carry a human crew to low Earth orbit, transfer the crew to a proper spaceship, and then wait in Earth orbit. Eventually the crew will come back to the ship and use it to reenter Earth's atmosphere. All manned entry and exit from space will be in these craft, unless a HAB is being launched from Earth by a heavy-lift rocket. For certain small, lightweight deliveries, SSTOs can also do runs, but it is not practical to carry large amounts of mass with this method.

Ice Mining/Terraforming

Celestial Propellant Package 2 (CPP-2): This is very similar to the CPP-1, except that it will have a large number of small tunnels melting ice and converting it to steam. The steam will be released on the surface of the object, out of the way of our mining equipment. A CPP-2 will be used both by a Corvus and by an MTESR to melt into and out of an ice object.

Tunneling Extra Surface Reactor (TESR): This is an ESR without the components that convert thermal energy from nuclear fuel into electrical energy. They will overheat under normal conditions, but the TESRs can only be active while tunneling through a large amount of ice to keep them cool. This will make a tunnel large enough to move a fusion device.

Micro Tunneling Extra Surface Reactor (Mtesr, pronounced 'em teser'): This device will be the same idea as a TESR except that it will be designed to make a large number of tunnels that are as small as possible. It will have a CPP-2 attached so that it can move the melted water/steam out of the tunnels and put it out of the way on the surface. This will be used for channeling energy from nuclear devices to break up objects along the line of our tunnels.

Corvus (CR): This is named after a boarding ramp with a hook on the end used by the Roman navy. It is a simple device that melts pieces of metal into an ice object, serving to anchor a man-made machine to an ice object with low or zero gravity. When reheated, the corvus detaches from the ice.

Track Assembly (TA): This is a system of tracks, an engine, and suspension. It will be set to drive a multiton ground vehicle around the surface of Ceres (designed with low Cerean gravity in mind).

TESR rig (TESR-R1 and TESR-R2): This is a large vehicle powered by an ESR. It will be able to extend a TESR into an asteroid. On Ceres, a TESR-R1 rig will have a track assembly attached to it. While on a zero-gravity asteroid, it will have a corvus instead. The TESR-R2 will work in zero- or low-gravity conditions (it will have a corvus for keeping it stable). It will push the TESR down by attaching interconnected support beams and extending them down through the ice. To pull the TESR up, it simply retracts the support beams. As the TESR melts through ice and turns it to steam, an insulated pipe will be attached to the support beams that will move the steam up to the surface (CPP-2). In space, steam will flash cool to ice (it will refreeze at the top of the object). On Ceres, the rig will be able to drive around the surface. In space, it can be moved around by a proto-ballista.

Support Beam (SB): There are multiple uses for support beams throughout this book. To be clear, we will design one standard piece of metal and use not only the same design but literally the same individual pieces (with different activities over decades). A support beam will be one hundred meters long, one centimeter thick, and ten centimeters wide. They will be made out of nickel-iron, because this is both strong and the most abundant nonmarketable material on asteroids. SBs will have interlocking grooves so they can connect to each other under low stress without welding (like a key and a lock). They will also have eyelets into which steel tethers can be attached.

Support Beam Stack (SBS-1 and SBS-2): This is a simple vehicle that holds support beams. It has a conveyor belt on it, which connects to the eyelets on the SBs. SBSs can be connected to each other to form a very long conveyor belt. They can also connect with all of the rig platforms. The rigs will grab support beams off the SBS and use them to push the Ice blade down through the ice. When pulling the ice blades up, the rig will pull the support beams up and put them on the conveyor belt (the reverse of the first process).

Standard Tether (ST): This is basically a chain made of nickel-iron or other cheap material. All vehicles will use the same size tethers, even the same tethers over several years.

Ice Blade (IB-1 and IB-2): This is a large piece of silicon, or thermal insulating material, with a narrow, metal strip at the bottom. These will be used to melt blast seems into Ceres. Once the sides of a megapili have been melted out, a fusion device will be better able to push it into space. The Ice Blades will use CPP-2s to transfer steam from the sides of the megapili up to the surface, where it will refreeze out of the way. At high settings, they can melt through metal or rock, but the goal is to only melt ice.

Ice Blade Rig (IB-R1 and IB-R2): This is the exhaust and support system for an ice blade. This will actually push the blade the full distance down into Ceres (using support beams to extend and retract). The support beams will be fed to the ice blade rigs from the support beam conveyor stack. It will also regulate the steam coming up and dispose of it as refrozen ice out of the way. This will have an ESR to produce electricity and either a corvus or track assembly based on the gravity of it's location.

Aluminum Reflector: This is an aluminum reflector made of huge pieces of super thin aluminum foil. AR-1s will be square in shape, sixty kilometers by sixty kilometers. We can make ad hoc shapes as needed (AR-2, 3, etc.) They will be designed with grips that match PBs, STs, and SBs.

Mars SSTO (MS-1): This will be as similar as we can make it to the ES-1, but it will be specifically designed to carry humans from the Martian surface to Martian orbit. Since there will be less gravity, atmosphere, and available maintenance on Mars, the design will have some very different features. It will have extra fuel-manufacturing capability to combine the CO_2 in the Martian atmosphere with stored hydrogen to make rocket fuel. MS-1s will go in and out of Mars orbit then rendezvous with a Hab for the ride back to Earth. In Earth orbit they will rendezvous with an Earth SSTO.

Venus SSTO (VS-1): This one will be as similar as possible to the MS-1 (possibly the same craft but with a modified engine, since the Venus missions will be after the Mars ones). Since we're making the planet's earth like, very similar vehicles will be able to work on Earth, Venus and Mars.

Air Batteries (AB-1): This is a two-sided molecular sieve. When one side is open, it will take in oxygen atoms. When the other side is open, it will take in CO_2. This will be a simple way of transferring gas between a green station and a human-inhabited area.

Fusion Igniter (FI): This is a device capable of nuclear fusion, but designed to have components removed, rendering it inert. These will stay in the custody of military service members from the state contributing them. All of these devices will be made on Earth from existing atomic weapons and will be only a few hundred kilograms.

Fusion Secondary Device (FSD): This is a large cylindrical device that, when given energy by a fusion igniter, changes the size of the blast from five hundred kilotons to several thousand megatons. It is made of concentric layers of dense metal and fusionable material (lithium). These are useless without the FI—they are not even toxic. These devices will weigh approximately 710 tons (the specific number based on the blast size). They will be produced in space. The FI will plug into the FSD, like a key into a lock. Once they are connected, it is a fusion device.

Standard oxygen tank: We will generate oxygen in space from plants and from electrolysis of water. We will use standardized containers with different air settings. When a crew has used all of the oxygen in one container, they can adjust the molecular sieve on the container so it will only let in CO_2. Eventually they'll connect to a station with plants or electrolysis capability and swap out their containers. The only vehicle that will have an oxygen container is the HAB; otherwise it will only be on propellant stations or green stations. Propellant stations will make oxygen as a byproduct.

Standard hydrogen tank: Hydrogen will be the primary propellant used by the MT-1 and the NTRO. Hydrogen will be made by breaking down water (H_2O) into its subcomponents (2H and O). Generally anything with an MT has a small hydrogen tank, and anything with an NTRO has a large hydrogen tank. Both size tanks will work on either vehicle. This will allow us to adjust our mass, energy, and resources better, based on the individual mission. The first few vehicles will launch with the tanks filled on Earth. As the tanks become empty, they will stay in space. Eventually we'll be able to fill these tanks in space through electrolysis.

Table 1: Subcomponents

	NTRO	ESR	MT	SPU	GRS	OSD	CPP	RTS	MA-1	MA-2	CR	TA
Kata	1	1	1	1	1	1	1	1				
Ballista	1	1	1	1						1		
Smelter		1	1	1			1		1			
Asteroid torch		1	1	1						1		
Proto-ballista		1	1	1	1				1			
GGP			1	1	1	1		1				
Martian balloon pod			1	1	1							
HAB				1								
TESR-R1		1		1								1
TESR-R2		1		1							1	
IB-R1		1		1								1
IB-R2		1									1	
SBS-1				1								1
SBS-2				1							1	

CURRENT WORDS EXPLAINED

Gravity well: Almost all of the energy that is used for space travel is expended to simply leave Earth. This applies (to a lesser or greater extent) to all large celestial objects. Once a vehicle is outside of Earth's gravitational pull, it requires very little energy to move immense distances inside of the solar system. This is analogous to being at the bottom of a well (on Earth). It requires a lot of energy and special equipment to climb out, but once you're at the top of the well, you can walk around very easily. Every planet, moon and large asteroid is at the bottom of its own gravity well.

Gravity assist: A gravity assist is the use of the gravity (equal to acceleration) of a planet or large body to increase the velocity of another body. The acceleration is used to speed the object up toward the more massive object, but it approaches at such an angle that it does not get pulled in but merely toward the massive object. This speeds up a space vehicle without causing a collision or even getting caught in the larger objects orbit.

Atomic bomb: Bombs based on the principal of fission, these are from one kiloton (North Korea's) to five hundred kilotons in yield.

Fusion or Hydrogen bomb: Bombs based on the principal of fusion, these have been detonated with yields as high as fifty-five megatons (fifty-five thousand kilotons). A device of this kind uses a normal atomic bomb as an igniter and then uses the energy from the first bomb to condense a secondary atomic fuel (usually lithium), making the explosion much bigger. There is no known size limit to this method of bomb manufacture, and the yields are very cheap megaton to dollar.

Nuclear thermal rocket: A rocket designed to use a nuclear reactor to superheat gas that is then directed out of the engine, generating thrust. Two substances are needed for an NTR, a nuclear substance to burn (emit energy) and a gaseous substance to propel. This type of rocket was tested successfully by NASA in 1973 and deemed space worthy. There were great advances made in NTR design during the Nuclear Rocket Program (Project Rover from 1955–1973). The math used in this book is based on the last of these designs (called the "solid core" rocket). The government snatched defeat from the mouth of victory by canceling the program after it succeeded in 1973. The most advanced design hypothesized (the gas core NTR) may be capable of speeds five times those of the solid core. The solid core is merely twice the speed of chemical propulsion we plan to use for the next 30 years.

Radioisotope thermoelectric generator (RTG): This is a simple device that generates electricity by harnessing heat moving between two metals. It requires no motors or magnets. This is frequently employed in modern spacecraft. These devices work best at a low level of intensity for a long time. The radio tag system (RTS) is basically an RTG with a small radio at the end. We will also use these to convert some energy out of spent nuclear fuel cells.

Variable Specific Impulse Magneto-Plasma Rocket (VASIMIR): This idea uses a propulsion design in which gas is superheated into plasma, and then the plasma is directed away from a spacecraft via electromagnet waves. This has the secondary effect of shielding the crew from radiation (the same effect as a Helmholtz coil). Hypothetically, this would result in twice as much velocity as a nuclear thermal rocket (four times existing chemical rockets). A former astronaut named

Dr. Franklin Chang-Diaz has been researching this. This system has never been built, but the design is reasonable.

Nuclear Salt Water Rocket (NSWR): This is a rocket design where the nuclear fuel is dissolved in water, and it is mixed in the right proportion to create a nuclear reaction immediately behind the engine. This design is similar to a liquid core NTR, but since uranium salt is used as a liquid, the physics is very different. This has never been built, but all of the physics works out. It is theoretically capable of incredible speed, though at a high rate of consumption for atomic fuel. This would be ideal for moving human crews quickly to the back of the solar system (particularly when a rich vein of fissile material is found on an asteroid). Machines would be sent on slower, cheaper transport.

I would like to point out that the chemical rockets we currently use are not drastically different from the ballistic missiles employed by the German military in the Second World War. Most practical attempts at innovation in space propulsion were abandoned in 1973.[6]

We will make an honest analysis of VASIMIR, NSWR, and even more advanced variants of nuclear thermal rockets. While the initial development cost would be high, this would save a fortune by cutting the travel time for manned missions. It also means we can mine more asteroids in less time, so less cost.

We can use cheaper and simpler nuclear thermal rockets to move pili. Simply pushing pili into the correct course (since objects in space have no friction) would start a multiyear trip. Human crews can be sent great distances with more advanced

---

6   While visiting the Johnson Space Center, my tour guide mentioned that the plan to go to Mars is still based on using chemical rockets in the 2030s. Nuclear rockets were finished in 1973. Twenty years from now, the US government plans to be sixty years behind.

propulsion to oversee the ore being removed and then sprint back to Earth while it's floating back.

Expeditions to Mars, Ceres, Iapetus, and even the Kuiper Belt would be much easier the faster our propulsion is. We will do the math for these systems and only build them if they are profitable. But this is a question for engineers, accountants, and shareholders, not governments. Do you think cars would have been permitted if we had a nationalized horse system?

The existing chemical rockets that are sitting in silos right now—ICBMs—will be used for carrying the Katas, proto-ballistae, and ballista-1s into low Earth orbit. While stock will be traded for these systems, they are already made and could be acquired without the direct use of cash. Once in space, all travel outside of Earth's atmosphere will be conducted by NTRO (at a minimum). The speed and mass of fuel requirements for an NTRO trump the benefits of chemical propellants. Even then, all the NTRO fuel has already been produced, and these are sitting in government warehouses with other atomic weapons components that no longer have any use. The NTROs will ship us around the oceans of space—we'll use chemical rowboats to come ashore.

NEW WORDS

While these are not pieces of equipment, there are several words that I will use for new things that will have to be created under my plan.

Gravitational breaking: To be clear, NASA has already done this multiple times, but for some reason this is referred to as a gravity assist. This naming issue is based on the physics definition of acceleration, meaning any change in velocity. When the word *acceleration* is used by most people, it is used in the sense of positive change to velocity. This is particularly the case in the automotive industry and for far too many high school physics teachers. If you told someone you accelerated your car when it crashed into a tree, they would get in a big argument with you; however, this is the correct use of the word in physics. (Sixty to zero miles per hour in one second is the same acceleration as zero to sixty in one second; the vector is just turned at 180 degrees.) In this book, I use the term *gravity assist* for gravity being used to add to the velocity of a vehicle or object; gravitational breaking is the same concept, but it is used to reduce the velocity of an object.

Blast seems: This is the general name for cutting an ice object free from a larger ice object. To facilitate the precise use of fusion devices, we will be cutting out the sides of megapili prior to extracting them. This will change a purely spherical explosive force into a dimensional force (more like a bullet being fired out of a gun).

Blast cut: This is the singular name for one specific blast seem made by one ice blade machine. Ice blades will be raised and lowered many times to make all of the blast cuts necessary to complete a blast seem.

Blast tunnels: This is the same concept as a blast seem. Blast seems are one centimeter wide, and they are only supposed to separate a piece of ice from its parent body. Blast tunnels must be several meters wide, since the fusion device has to move down them to the deepest point of the megapili we are removing.

Mtesr tunnel: Same idea as a blast tunnel but much smaller. A megapili will have one TESR tunnel from the surface to its center of mass (to emplace the fusion device). A megapili may have tens of thousands of Mtesr tunnels to direct the energy of the detonation from the center of mass to the surface.

Newtonian ballast: This idea is based on Newton's third law of motion: "for every action there is an equal and opposite reaction." The principle of thrust in aviation is based on this idea; air is propelled by an engine, the force of this air is backward, the opposite reaction pushes the aircraft forward. For the purposes of atomic mining, breaking off a piece of an asteroid and propelling it in one direction will push the rest of the object in the opposite direction. Usually this will mean some useful portion being sent toward the inner solar system and a less useful portion being sent toward Pluto. The part used to push against is Newtonian ballast.

Cherson: From the Greek word *chersonesus* ("peninsula"), a cherson is an extrusion from an asteroid. Only very large asteroids are spherical in shape, and even they still have boulders and slight differences in terrain. Any piece that sticks out of an asteroid will be easier to break off and mine. On Earth we might call these hills, mounds, knolls or boulders. Eros 433 is shaped like a 30 km wide shoe, so the use of geographic Earth terms will be confusing. Due to the high content of valuable metal on asteroids, we don't need to dig at all. Just cutting a cherson from the surface, or even moving an object held down only by gravity, could be done at a huge profit.

Pili: While asteroids are currently floating all over the solar system, this plan involves the manipulation or deconstruction of these asteroids by man. I use "pili" to refer to all man-made or man-diverted asteroids to distinguish them from natural asteroids. When a cherson is removed from an asteroid, it becomes a pili.

Megapili: For terraforming planets, huge pili will need to be made. Any pili that is larger than one kilometer in diameter is a megapili. These will never be used for profitable mining but only for terraforming. (They will be kept very far from Earth.)

Sunward: This is the component of velocity moving toward the sun. An object can be moving in its orbit at fourteen kilometers per second, but it could also have a sunward component of velocity at two kilometers per second. This word is also used to describe the side of an object that faces the sun.

Plutoward (pronounced "ploo-toward"): This is the same concept as sunward but in the opposite direction. I realize that once we pass Pluto, this name won't make as much sense, but it sounds better than "universward."

Onworld and Offworld: I'm sure I've heard these words in science-fiction novels, though I cannot identify the source of them. Almost every object that any human has ever owned was onworld. Since all objects meet this description, there is no point in creating a separate adjective for it (no one specifies that one particular glove is made of matter). When mining is moved into space, there will be vast amounts of vehicles, people, products, and raw materials moving back and forth between Earth, space, and the other planets. There is an immense cost in moving something out of a gravity well. FedEx will have to track things in the solar system, not just on this tiny planet. It takes a lot of time to say: "Your package/friend/boss/raw materials/ Chinese takeout (it could happen) has arrived on this planet." Saying "its onworld" is much faster. Americans will recognize this phrase in the same vein as "offshore accounts" or "offshore drilling" (meaning not in the continental United States).

Litho-Glacis line: This is the name I am giving for the dividing line on a stellar body where the frozen volatile material stops and solid rock or metal begins. On Ceres, we know that this line is about one hundred kilometers down (though we don't know if it is a clear line or more of a general zone). There are many asteroids in the Cybele family and the Thermatid group that also may have ice on them. Determining where the ice ends and the rock starts will be very important. The definition of a comet is that it has ice on the outside (the entire object could be ice). Some bodies may have a secondary LG line (where water ice ends and where CO_2 ice begins).

PROPOSED WORD CHANGES

As we venture farther and farther out into space, our vocabulary will have to be as scientific and technical as we are. Before Galileo, people had a relatively narrow vision of our place in the cosmos. "The sun" and "the moon" were proper enough singular names (mankind did not realize other worlds had moons or that the Milky Way galaxy had 800 billion suns). Over the course of writing this book, I had to sift through information on about two hundred moons. I've taken to writing out "Earth's moon" to not get it confused with Phobos, Deimos, Iapetus, Ganymede, Titan, and so forth. Imagine how embarrassing it would be for someone living on Mars to get a low price on a ticket to the moon, only to realize he bought a ticket to Phobos (which Martian colonists would logically call the moon, since it's *their* moon). This frequently occurs with street names in cities or common town names in different states. We are only worried about one sun so far, but it may not be long until we have a multisolar civilization as well as a multiplanetary one.

Copernicus made his observations in an era when Roman numerals were still being used. Try writing the distance between here and the Andromeda galaxy with those! But now that our horizons have been deepened by optics and technology, I recommend the adoption (already frequently used in science fiction) of referring to our own immediate sun and moon by the Latin names of Sol and Luna. Even if this weren't commonly done on Earth, for people living and working in space, it would be very useful.

If you only knew one guy named Brad, Brad would be a sufficient name. If every male on this planet were named Brad, perhaps you'd have to give them all nicknames or use their middle names. This will not be part of the Space Corporation charter, but it was a constant problem I encountered when writing this book.

Speaking of this, as a space-faring society, we have really not given a lot of thought to the definitions and names we use for moons. Whenever the word *moon* is used, the idea is a substantial object ("I'll give you the moon"). Earth's moon is 3,474 kilometers in diameter. The moons of Mars, on the other hand, are twelve kilometers (Deimos) and twenty-two kilometers (Phobos). If I built a car that was less than 1 percent the size of a normal car, we would call it a "toy car" or "baby car." I have seen the word *moonlet* used in some literature to convey a lunar object of diminutive mass. I would propose setting a fifty-kilometer diameter size requirement for an object to be called a moon. Most of the moons of Saturn are smaller than the fifty largest asteroids in the main belt. The gas giant pods will find hundreds of moon-like objects we don't have on record now.

GPS will also eventually need to get a name adjustment. On Earth, this is the global positioning system. While I'm not going to sue or bother anyone for this wildly arrogant name (there are hundreds of globes in the solar system!), I will specifically call the comparable systems set up on Mars and Venus the "Martian positioning system" and the "Venusian positioning system" to avoid confusion. Hopefully in the future the system for Earth can be called the geo-positioning system ("geo" is Greek for Earth). This is very important since the ore-bundle parachute assembly will tie into the positioning system on whichever planet it is operating near. Minor differences in the parachute and the ablative shielding will have to be factored in for local atmosphere and gravity.

None of these word changes will be in the Space Corporation charter, but I'd be embarrassed were mankind to gesticulate and grunt at anything more complex than a sharp rock whilst we zip through the cosmos on our nuclear-powered starships.

CHAPTER 3

Reconnaissance

Recycling the Cold War

A large number of intercontinental ballistic missiles were amassed by the United States and the Soviet Union during the Cold War. Some have been accumulated by a few other states.

Most of these weapons have aged and were disassembled, destroyed, or fired into oceans or deserts without warheads. As far as weapons currently possessed by Russia, their main system is the R36 rocket, with a throw weight (mass to low Earth orbit) of 8,800 kilograms. The largest American heavy-lift ballistic missile is the UGM-113, with a throw weight of 2,800 kilograms. There are many different missiles on earth but I will focus on these because of their abundance and carrying capacity.

While this is not the rosy vision of peace that is usually conveyed to the public, the space race was a thinly veiled part of the nuclear arms race. An intercontinental ballistic missile works by flying a weapon into space and then flying through space at immense speed; it reenters the atmosphere and comes down over something one wants to blow up.

Placing Sputnik in orbit was comparable to the Russians (then Soviets) firing a machine gun over the head of a next-door neighbor and calling it "atmosphere exploration." Sputnik was mounted on the same rocket that Soviet atomic weapons were mounted on. Putting Sputnik into Earth's orbit was as difficult as planting Sputnik into the lawn of the White House at seven miles per second.

Basically I propose that Russia trade eight R36s and America trade eight UGM-113s (with all warheads removed) in exchange for stock shares in Space Corporation. These can be the oldest eight systems that each state controls to minimize the effect on that nation's rocket arsenal.

I spent five and a half years as an air defense officer in the US Army, and I trained on and against rocket systems. I can assure everyone that these rockets age, just like eggs; once they go bad, they're destroyed, which alone is an expensive task. The vast majority of rockets created during the Cold War have already been peacefully decommissioned. Anything capable of flying from the United States to Russia, or vice versa, can only do so by traveling through space. We can remove the warheads (and violent intent) from these rockets and put them to very profitable peaceful use.

If Russia and the United States were comfortable with this arrangement, the more rockets they turn over, the better. I've built the plan for the reconnaissance phase at about 1 percent of the American and Russian arsenals. In the forty years it will take to implement this plan, all 1,400 ballistic missiles owned by the United States and Russia will have been taken offline by their government anyway. Why not exchange them to Space Corporation once they hit their last year of service? We've already thrown away thousands of these rockets as they've gone obsolete in the last forty years.

These systems are built into fortified silos so they will never physically leave their government-controlled areas. A team from each nation's armed forces would remove the atomic warhead, and afterward a Space Corporation team would add the new Space Corporation equipment.

The United States successfully removed the warheads from fourteen Titan II missiles and used the rockets themselves to fly satellites into orbit. This bizarre experiment with government efficiency was promptly abandoned, but it does show that the capability is there.

Russia would gain more from this than the United States would, since their systems are older and larger and their economy is smaller. The US military has an expensive facility just for taking apart nuclear warheads, not counting the rockets, which are literally tons of super-flammable material. So with sixteen aging ICBMs, we shall move our Katas into space.

To be clear, America can provide all sixteen of these rockets simply by contributing the oldest 2 percent of our ballistic missiles, but it would be both polite and profitable to invite Russia and all the other democratic states of the world to help out.

I've mostly dealt with the United States and Russia since they have the most weapons, but France, the United Kingdom, India, and Israel will be invited to contribute missiles under the same terms. Pakistan, for example, has had civilian rule since 2008, so in 2018 they will be invited to join. Once other states have peaceful democratic elections for ten years, they will also be invited. This could be the most profitable—actually the only profitable—disarmament treaty in human history.

RECONNAISSANCE AND PROFITS

A key assumption of the things you must know before you solve a problem is the things you don't know! We have very sketchy data about the solar system at the level of detail needed for profitable mining. What quantity of rare earth metals are on the one thousand asteroids closest to Earth? What is the concentration of uranium, gold, platinum or any other marketable material on these asteroids? Are there asteroids completely or primarily made of ice? What's the percentage of ice, and how deep is it in relation to quantities of heavy metals (and which metals)? Which Kuiper Belt objects are of the best composition and orbit for a Mars impact? Which objects are of the best mass, composition, and orbit to be built into LaGrange objects and Helmholtz coils for Mars and Venus? Which largely ice asteroids with valuable material are in the best orbits for a Venus or Mars impact?

While we have learned a great deal from modeling the gravity characteristics of these objects, everything starts to get pretty dim once we get into the level of detail necessary for mining or terraforming. There is no other way to get specific information than to fly Katas out to these locations.

Since this is a simple, profitable plan for going into space, we will not be building small probes with delicate landing gear. All of the Katas will accelerate to full speed after they are carried into space. Radar waves travel at the speed of light. We frequently take radar signals from objects that are already in flight (for example, supersonic fighter jets and even missiles themselves are equipped with radar). Each Kata will fly close to its individual target, scan it by radar and GRS, and then make a slight course correction (using the target's gravity to turn and accelerate).

As a Kata is approaching the vicinity of Pluto, Eris, or Sedna at eighty thousand kilometers per hour, it will fly just above the height for a stable orbit. All of that gravitational energy will still pull on the Kata as it passes, slowing or stopping it (gravitational breaking). The Kata can then use its MT to make minor adjustments or alter its specific course. Even if the Kata is not near one of the large objects in the Kuiper Belt, it can still fly very close to several small objects, each of which will slow the Kata slightly with its gravitational pull. This will save us maintenance cost, propellant, and fuel.

I have written this plan to be a general strategy for colonizing the solar system. Specific targets and time tables will of course be adjusted based on what the Katas, Gas Giant Pods, and Martian Balloon Pods find on their survey. We will start by launching a total of sixteen Katas over a two-year period to the ninety-six targets on table 2.

Table 2. Kata reconnaissance targets and payloads

Kata#	Target 1 Near Earth or Venus	Target 2 Asteroids	Target 3 Asteroids	Target 4 Comets	Target 5 (Gas Giant Moon)	Kuiper Belt	Payload
Kata1	(Earth's) Moon	1999 UJ₇ (ML4)	Ceres*	2060 Chiron	Phoebe (Saturn)	KB Area 1	GGP?
Kata2	Apophis 99942	5261 Eureka (ML 5)	3 Juno	174 P Echecklus	Proteus (Neptune)	KB Area 2	GGP
Kata3	2005 GO₂₁	2007 NS2 (ML5)	7 Iris	8405 Asbolus	Miranda (Uranus)	KB Area 3	GGP
Kata4	3554 Ammun	2001 DH47 (ML5)	24 Themis*	52872 Okyrhoe	Adrasta (Jupiter)	KB Area 4	GGP
Kata5	3753 Cruithne	2011 SC191 (ML5)	87 Sylvia*	596 Scheila*	Iapetus (Saturn)	KB Area 1	PB
Kata6	5381 Sekhmet	15 Eunomia	65 Cybele*	10 Hygea*	Nereid (Neptune)	KB Area 2	PB
Kata7	1989 VA	2 Pallas	107 Camilla*	90 Antiope	Margaret (Uranus)	KB Area 3	PB
Kata8	1990 VA	511 Davida	121 Hermione*	4 Vesta	Amalthea (Jupiter)	KB Area 4	PB

Kata9	1994 TF$_2$	Phobos (Mars)	76 Freia*	2 Pallas	Hyperion (Saturn)	KB Area 1	MBP, CMS-1
Kata10	1994 WR$_{12}$	Demos (Mars)	790 Pretoria*	468 Lina	Psamathe (Neptune)	KB Area 2	MBP, CMS-1
Kata11	1998 DK$_{36}$	Eros 433	*566 Stereoskopa	526 Jena	Titania (Uranus	KB Area 3	MBP, CMS-1
Kata12	1998 UP$_1$	16 Psyche (M)	704 Interamnia	846 Lipperta	Metis (Jupiter)	KB Area 4	MBP, CMS-1
Kata13	1998 WT$_{24}$	21 Lutetia (M)	31 Euphrosyne	167 Urda	Irijiraq (Saturn)	KB Area 1	MBP, CMS-1
Kata14	1998 XB	216 Kleopatra (M)	532 Herculina	46 Hestia	Neso (Neptune)	KB Area 2	MBP, CMS-1
Kata15	1999 KW$_4$	22 Kalliope	944 Hidalgo	8 Flora	Oberon (Uranus)	KB Area 3	MBP, CMS-1
Kata16	2000 BD$_{19}$	52 Europa	6 Hebe	152 Altala	Thebe (Jupiter)	KB Area 4	MBP, CMS-1

7   GGP: Gas Giant Pod, PB: Proto-ballista,
MBP: Martian Balloon Pod, CMS: Communication Satellite, *Object thought to have Water, (M) Metallic asteroid
() denotes planet that moon orbits. Phobos orbits (Mars)
ML4 & 5 are at the Martian LaGrange point 4 & Martian LaGrange point 5.

These sixteen Katas are organized by their payloads/targets. These targets were selected for either being Earth-Venus crossers, large in size, comets, or planets and moons in their own right. The area of the Kuiper Belt scanned by each probe will be the part closest to the gas giant that probe visited. KB Area 1 will be the part of the Kuiper Belt nearest to Saturn, KB Area 2 the section nearest to Jupiter, KB Area 3 section nearest to Uranus, KB Area 4 section nearest to Neptune.

This is a general guideline for useful targets; the precise list will depend on the year of the launch and orbital alignment of the various targets. There are sixteen reconnaissance missions that I have placed in six different phases for naming and clarity. It is not part of the plan that they be launched in numerical order. There are also millions of objects in the solar system, and each Kata could be assigned one hundred thousand objects, but here are the ones that seem worth our time.

I'm assuming that it will take four years to reach the Kuiper Belt (the front of the belt is 2.7 billion miles from earth, the back of the belt is 4.5 billion miles from Earth). This is based on twice the speed of the New Horizons probe launched to Pluto, which used a specific impulse of 450 seconds (less than half of what the nuclear thermal rocket tested by NASA attained).

GAS GIANT SCAN

Katas one through four will be launched from Earth with a gas giant pod attached to each. The Katas have an optical spectrum device, a gravimetric radar scanner to find compositional data, and a radio tag system to mark the location of interesting asteroids.

All of the targets on my list have known positions and will not need radio tags. The radio tags will be needed if the Katas chance upon something valuable or for an object that is very small in size. The radio tags on the gas giant pods will be important for marking tiny objects amidst the multitude of things orbiting the gas giants. If a one-cubic-kilometer piece of methane was found around Uranus or an eight-hundred-cubic-meter piece of gold found around Neptune, it would be really useful knowing exactly which asteroid it is out of five hundred thousand.

There is also not much need for the Katas to take pictures in the visible light spectrum. Some of them can be seen fine from Earth while others have no light in the visual spectrum to capture anyway. The main tools used by the Katas inside of the Kuiper Belt will be infrared cameras, radar, and the GRS. These systems will tell us where objects are, how they are shaped, and what their density is.

After each Kata has scanned the first five targets it is assigned, it will arrive in the Kuiper Belt. The gas giants orbit the sun faster than the Kuiper belt objects (Jupiter at eleven years, Saturn at twenty-nine, Uranus at eighty-four, and Neptune at 164). The objects in the Kuiper Belt orbit the sun at various different speeds, but a reasonable average is the orbit for Pluto, 247 years. With four different GGPs arrayed throughout the solar system, we will have plenty of retransmission options for the probes in the belt.

The radar will be used frequently to help the Kata see where it's going, the gravimeter does not consume any power (it literally measures the gravity experienced from a nearby object). Each of the Cybele asteroids has thousands of small ice asteroids nearby. When the Kata identifies an asteroid that is mostly ice, it will attach a radio tag to it. We can have each Kata pick one Earth-use ice asteroid (there are six Katas going to the Cybele group, and one for 121 Themis). This data will be beamed back to Earth after each scan. We can factor this information into other Kata missions (more scouts to an area of interest, fewer to an area that appears empty). The Katas will be programmed to fly close enough to scan an object and measure its gravity, but they will not stop or even slow down. We can use the gravity of some objects to adjust their flight path if we want to conserve fuel and engine life.

The gas giant pods will simply be carried freight until each Kata approaches its respective gas giant. Once the Kata lines up in the right trajectory, the GGP will be released into the orbit of its target gas giant. The GGP will move into a stable orbit. From there it will spend months collecting all of the data it can about that planet's orbital debris and moons. After visually cataloguing objects, the GGP will use its maneuver thrusters to widen its search area. It will move to a different orbit and stay there for a few months conducting a new set of scans. In this way the entire orbit of a gas giant can be scanned over a 2 year period.

One of the key things we're looking for among the gas giants is methane. Methane is seventy-two times more potent of a greenhouse gas than CO_2. Most methane is in the Kuiper Belt, though there are small pockets closer (by Uranus for example).

The distance within the Kuiper Belt is also gargantuan— the near side of the belt is thirty astronomical units (one astronomical unit is 150 million km[8]), and the far side is fifty astronomical units. By sending sixteen Katas to the best spots we know of in the Kuiper Belt and ensuring the Katas have as much artificial intelligence as a Predator drone, we can be reasonably certain that they will find the objects we want and radio back the compositions and locations.

A few large objects (one hundred kilometers across) have been observed in the Kuiper Belt, but it's reasonable that there are hundreds of small objects for every large one. We're looking

---

8   The exact number is 149,597,870,700 meters.

for a methane object with a four-kilometer diameter, with as sunward a path as we can find. Hopefully we'll find and catalog hundreds of objects, selecting the four best. We might need multiple runs to make Mars exactly earthlike, but we can make a lot of progress with four large methane impactors.

The first Kata will go to Saturn. Jupiter has very strong radiation that could interfere with the Kata or the GGP's electromagnetic equipment. We may even want to alter the GGP earmarked for Jupiter and have the Kata release it on a coasting trajectory from a great distance. The second Kata will go to Neptune. The third Kata will go to Uranus. Each Kata will release its gas giant pod, scan one or two moons (without slowing down) then perform a gravity assist around its target planet to accelerate to the nearest part of the Kuiper belt. The other twelve Katas will generally be directed into clusters of four (so three Katas will join each of the original four in the Kuiper Belt). Hopefully, these first four Katas will find suitable impact targets and then go on to amass more data.

Once the first four Katas arrive in the Kuiper Belt, they will search for potential Mars impactors. The objects that meet our mass, orbit, and composition characteristics will have radio tags attached to them. A year after the first Katas arrive in the belt, all of the information will be studied, and four KBOs will be selected by our analysts on earth.

PROTO-BALLISTAE

The next four Katas (five to eight) will be launched on UGM-113 rockets. These rockets will carry a Kata and a proto-ballistae (PB-1). The Katas themselves will proceed to scan their specific targets; the PB-1s will remain in Earth orbit. The PB-1s will start gathering the large amount of human-created debris in Earth's orbit left over from previous space missions. This is not directly tied to the reconnaissance operation, but it will be mentioned in chapter 4. The PB-1s will have a lot of work to do; the sooner they get started the better.

We will launch the first ballistae (BB-1) in year five so the PB-1s launched in year four will allow us to test many of the different systems while also gathering up potentially dangerous debris. Mining is the first goal of Space Corporation, but the PB-1s can do a lot of useful work while waiting for mining expeditions.

The proto-ballistae will have many different uses for mining and terraforming, but clearing up debris in orbit will be a useful diagnostic. In the long term, removing debris from orbit will make all of our operations much simpler.

SCANNING MARS

The next series of Katas are nine to sixteen. These will be launched with the Russian R36 rocket. All of the Katas themselves will be identical, but the attached equipment for these eight systems will be Mars balloon pods (MBPs) and communication satellites (CMS-1s).

This series of Katas will push an MBP and a CMS-1 with them while scanning objects en route to Mars orbit. Near Mars, the Kata will detach the CMS into Mars orbit and the MBP directly toward Mars.

The MBP will descend into the Martian atmosphere, discard its heat shield, inflate a large balloon from stored compressed air tanks, and then discard the air tanks and pumps. It will then float around Mars for two years. We'll focus these eight balloons on the Martian equator and the Martian poles. The first human landings will be along the equator, and the asteroid impacts will all be at the poles. Of course, we'll try to scan the entire planet if we can.

Roughly every two Earth years (one Martian year), there is a dust storm that covers the entire planet. A balloon may be able to fly in this, but it would be harder to control and bring in lower quality data. We'll plan all of the balloon missions to arrive just at the end of the Martian summer so they will have the best chance of one and a half Earth years of good weather.

We'll have to decide whether we want a safe high scan or a dangerous but more detailed low scan. We can start scanning at a high altitude then lower the altitude with time. The high scan will give us a lot of general information. We'll lose the probes after 2 years anyway, so we can keep lowering them for better data until they crash or have controlled landings from power loss. Doing this in the Martian atmosphere will get us data at least for the Martian crust, and this will be vital when picking sites for asteroid impacts and human bases. The spots with the most frozen water or frozen CO_2 will be targets for asteroid impacts. Places with potentially good soil or a good concentration of expensive metals will be targets for landing sites.

The balloons will be powered by an SPU. They will be designed to float around Mars for two years, broadcasting the results of their gravimeter. We'll plan on the SPU being able to sustain this for two years (the time between dust storms). At the end of that time, the balloon will have a controlled landing near the future human settlement area (just south of the Valles Marineris). If the wind patterns have moved it completely on the opposite side of the planet, we will at least try to bring it down gently and on a piece of ground that is above the future sea level.

While the small generator on the probe will be out of power, it will have many other useful components in relatively good condition. These generators (SPUs) will be the same ones eventually used for the Mars landing sites, and even old equipment will be a valuable source of spare parts or backup systems. Any GRS systems that survive landing when their balloons run out of power can eventually be used in future surface prospecting.

Maneuvering these balloons around Mars will be difficult (we'll mostly rely on wind currents), but we can put them into the atmosphere at very different points to ensure a wide scanning area.

The easiest way to move something around a planet is to move it around *before it's on the planet.* This was why ballistic missiles were invented in the first place: the fastest way to get something from America to Russia is by taking the shortcut through outer space.

The communications satellites in orbit will allow us to get grid coordinates for all of the gravimeter scans the balloons make, and send this data back to earth to be stored and studied. Since a huge amount of data will be collected, we will also be broadcasting back to Earth as quickly as we can. These satellites will be built to last a long time so they will be there through all our shenanigans.

RECONNAISSANCE CONCLUSION

Now we have sixteen Katas in the Kuiper Belt, and we've scanned more than one hundred large mining and terraforming objects and thousands of smaller objects with our GGPs. We've brought four small robots to Earth orbit to collect space debris, dropped eight balloon probes on Mars, and dropped one probe into each of the orbits of all of the gas giants. We will now be able to refine our plan on where to go to make profit, to find material for terraforming, to make equipment, and to cut our operating costs in space.

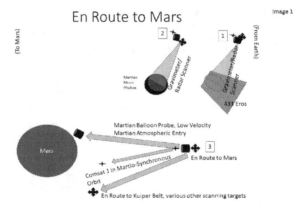

En Route to Mars

Image 1

CHAPTER 4

Infrastructure

Recycling Space Junk

There are growing concerns about the amount of space debris that we have in orbit around Earth. In 1978, NASA scientist Donald Kessler proposed a scenario where space debris would become so common it would collide with other space debris in a runaway collision event. An example of this can be found in the collision of the functional US satellite Iridium 33 and the derelict Russian satellite Kosmos 2251 in 2009. More than 1,700 pieces of debris formed from these two satellites, and there's plenty more where that came from.

There have been multiple small impacts with American space shuttles. In addition, there is considerable effort in moving space vehicles to avoid passing debris. The US Army Space Command spends an enormous amount of time and money tracking all of these pieces traveling twenty-five times the speed of sound. If you don't have a clear mental picture of this, imagine paying someone to count all the pieces of litter flying through the air during a tornado.

Instead of viewing the growing amount of space debris as a problem, we should think of this as the back of a solution. Highly sophisticated equipment for Space Corporation will have to be made on Earth, but a lot of high-mass objects can be easily melted down from debris in orbit. Many of the components floating in space are final stages of rockets that carried their payloads out of the atmosphere. We will not be using the same type of rockets that were used fifty years ago, but we will be using the same type of metal for our rockets.

We will eventually use our celestial smelter to melt ore for asteroid mining, and what better way to test it than by melting some objects of known composition? This will work exactly the way that scrap iron does on Earth, but we will be doing it for bulk components in space. The habitation module would be much safer with a two-thousand-kilogram metallic shield wrapped around it. We will build the shield in space and attach it once the HAB breaches the atmosphere. We can build a second layer for the shield when we eventually send a HAB toward Venus (more radiation means a shield of more thickness). These shields will just be a big hunk of metal. It would not have been complex for them to have been made by the Greeks in the eighth century BC; they just would not have been able to get them into space.

There are an infinite number of possibilities with this idea. Once we start making useful Space Corporation components in space, we can also sell this service to other companies on Earth for any material that we don't need. If someone wants to build a hotel, mansion, or Jacuzzi in Earth orbit, no one else will have thirty thousand kilograms of construction material and a smelter *already in space*. How much money would you pay to use a zero-gravity Jacuzzi?

SPACE-DEBRIS RECYCLING

The first things launched into space will all be made on Earth and launched in the fourth year of the Space Corporation. There is no fundamental difference in melting a big satellite or a small asteroid. The satellites don't have any metal that would be cost-effective to return to Earth. They do have many metals that would be useful to test our smelter and to use as feedstock for making our own cheap components in space.

Some of the equipment that we need can be developed over several years, and we can plan from the start to use melted space junk for bulky simple parts. This space junk will be purchased in exchange for stock from its original owners. There are currently forty-six derelict satellites in Earth orbit, which would be easy to scrap, as well as productive to move out of the way.

There are thousands of other small pieces that we will collect, particularly the ones with orbits near functioning space vehicles. We can even charge money for this service, since moving satellites and space stations around debris is currently a large expense. There are also many useful derelict objects orbiting the sun near Earth's orbit.

Some of these pieces could be salvaged and used intact. There are currently four Saturn V upper-stage fuel tanks in heliocentric orbit. The US space station Skylab, with a crew of three astronauts, was made of one such piece. Similarly, the entire International Space Station (ISS), minus a few separate compartments, is scheduled to be scrapped in 2024.

When all of these pieces are added together, it comes to approximately six hundred thousand kilograms of large usable pieces in stable orbits. This is a junkyard beyond mankind's wildest dreams; this equipment is at the top of the gravity well. At the current NASA price to low Earth orbit this debris is worth US$110 billion in launch costs. If we offer five thousand Space

Corporation shares at US$5 apiece, we'll have an incentive of $109.999975 billion to clean it up. Under existing laws, businesses or governments that own space debris are legally accountable if that debris crashes into anyone or anything. Most states and businesses would love to be rid of these liabilities.

The proto-ballistae would be more than capable of grabbing these satellites, even disassembling them and feeding the parts into a smelter. We will collect these pieces in order from largest to smallest, and also in the order of most dangerous to least dangerous relative to functional space vehicles.

Besides the large and intact derelicts, there are nineteen thousand pieces that are larger than one kilogram. There are five hundred thousand pieces larger than 0.1 kilogram. There are tens of millions of objects smaller than that. At each level, the impact hazard goes down, as does the profitability of harvesting them, but we can do much better than nothing!

We will plan the orbits of our debris and our smelter so that the smelter can easily intercept and melt down our orbital junkyards. Deciding what to smelter, what to use it for, and when to do it will be very complicated. Having a billion dollars of material in low Earth orbit is a problem I would love to have.

PROPELLANT STATION 1

Nuclear thermal rockets rely on a small amount of uranium as fuel for a nuclear reaction and a large amount of liquid hydrogen as a propellant. Many other substances can be used, but liquid hydrogen works best. Since there is no friction in space, travel in space will quickly surpass travel on Earth in price per mass. Shipping an object from Mars orbit to a smelter in Nevada will be cheaper than trucking something from Wyoming to Nevada.

The uranium generates heat, and once the hydrogen is heated, it expands. This expansion can be channeled out of a specific nozzle to generate thrust. We only need a tiny amount of uranium for this, but we need much more hydrogen. Producing hydrogen in space (since hydrogen literally makes up most of the universe) is an easy way to lower the cost of operations.

In year four, we will launch several Katas that will fly through the asteroid belt. We will send them into areas thought to contain small ice asteroids (there are thousands in the vicinity of the Cybele group). By year five, we will have the first information from the Katas confirming what they have found. We will then launch a ballista with full propellant, which will fly to the main asteroid belt, grab one ice asteroid of approximately ten thousand tons, cover it with Mylar so it doesn't melt, and bring it back to Earth orbit. If an ice object this small were to accidentally enter Earth's atmosphere, it would melt from the immense heat of reentry.

Electrolysis is the simple method of running an electric current through water, which then splits water molecules into oxygen and hydrogen. Solar panels are a primary component of every vehicle that has been put in space, even the ones that no longer work. Electrical wiring is very easy to connect together (unlike air or liquid piping).

As of this writing, the ISS will be taken offline and disassembled in 2024. For some reason, it was intentionally placed in an unstable orbit, and it degrades by two kilometers every month. This requires a large amount of money to keep pushing it back up to a stable orbit (an effect of physicists working without businessmen).

Without massive contributions of cash in the way of fuel launched out of the gravity well, the ISS will eventually crash to Earth or be dismantled. If the ISS were sold intact to Space Corporation in exchange for stock, it could be pushed into a stable orbit and converted for use as an unmanned hydrogen reactor. The power cells on the ISS produce 83.6 kilowatts per day, while only being exposed to the sun 50 percent of the time (again, an effect of its low orbit). This output would effectively double to 167 kilowatts per day once the orbit was stabilized, enough to split four kilograms of hydrogen while making thirty-two kilograms of oxygen. We'll bring the water ice in from asteroids in the main belt. The oxygen will be harvested using air batteries. The liquid hydrogen will be cooled by the shade of the sun, and the hydrogen gas will condense to a liquid on the other side of a molecular sieve. We can place the water in a translucent container that keeps it warm enough to be liquid but as cool as possible (the electrolysis reaction is more efficient at cold nonfreezing temperatures).

The hydrogen will be filtered out from the water, and this will be the propellant that we will use to travel in space. Both the NTRO and the MT will be designed to use hydrogen as a propellant, with standard fuel pellets as the actual source of energy.

We will plan for mining expeditions to take place at three-month intervals (to correspond with food production on Green Station 1). At three months, 360 kilograms of hydrogen fuel will be available for propellant, and 2,880 kilograms of oxygen for the crew to breathe. This is eight times what a four-person crew will need to breathe for three months. We can even use oxygen as propellant; though it is not as efficient, it will be very abundant.

Our oxygen storage tanks will fit 400 kilograms of oxygen, enough for a three-month trip to Mars for a four-person crew, plus 10 percent. The smallest hydrogen tanks will be 50 kilograms (proportionate to the oxygen of a water molecule 1:8). We can make larger tanks as needed, but these will be our building blocks.

All of these hydrogen and oxygen tanks will remain in space and fit to the hull externally (they'll be removed if the HAB is landing on a planet). The oxygen tank can also be plugged in as a propellant tank. Space Corporation will also use air batteries to store and disperse air. These will be smaller and kept inside of the HAB or an SSTO.

We'll make the hydrogen tanks in several different sizes, based on function. All sizes will have the same connector pieces. We'll be able to adjust the size of the fuel tanks based on the specific mission or specific equipment it is attached to. This will make cycling all of the tanks through easier and allow us to adjust for duration (more propellant) versus maneuverability (smaller tanks).

Of course, we'll have to change the name from the ISS to something more corporate sounding (Space Station Nike? We're obviously comfortable with corporate sponsorship.) I'll let the legal department figure that one out. For now, let's call it Propellant Station 1.

Propellant Station 1 will be our first source of propellant in Earth orbit, but there are plenty of smaller pieces of debris, most of which have solar panels. It would be simple to attach these pieces on a large, thin frame and use them to increase the electricity going into Propellant Station 1. We can either augment the number of panels on Propellant Station 1 or make more stations. We will be melting these objects to make radiation shielding for human crews and Green Station 1. We can produce even more fuel using nuclear reactors in orbit, which I will touch on later.

We can even save cost and move some hydrogen fuel from Propellant Station 1 down to Low Earth Orbit (LEO) in small fuel tanks. This will allow vehicles to launch from Earth into low-altitude orbits, pick up a hydrogen fuel tank intentionally placed at their altitude, and then use this new source of fuel to continue flying to a stable altitude. Overall, making propellant in space will cut the cost of space missions from billions to millions, and it is at the core of a successful settlement of space.

NUCLEAR PROPELLANT PRODUCTION

Using Propellant Station 1 with its existing amount of solar cells will produce four kilograms of hydrogen and thirty-two kilograms of oxygen per day. Space Corporation will not be limited to this amount; this is just a cheap, long-term method of production.

We have a more expensive, though much faster, machine that can break down water: the celestial smelter. Just as ore can be exposed to heat in order to separate different elements by melting point, enough heat can also break water into hydrogen and oxygen. One kilogram of U235 can produce 8.64E+13 joules of energy. This is enough for thermolysis on 3,440,000,000 kilograms of water, which would yield 430,000,000 kilograms of hydrogen and 3,000,000,000 kilograms of oxygen. In the beginning, we will have very little demand for oxygen, so we may just dump it into space until we build enough storage containers for it.

Besides using thermolysis on water to create hydrogen propellant, we can also build a small electric generator that attaches to the smelter. We can heat steam inside of the smelter and then direct that steam into a generator to create electricity. This gives us the large volume of energy available with nuclear power, but we don't have to deal with as much heat as with thermolysis. One kilogram of U235 can electrolyze 573,333 kilograms of hydrogen. A very simple nuclear reactor could be attached to a smelter and harness the energy from heated steam to produce electricity. This would probably generate less wear on our equipment than using thermolysis, but both options are open to us.

The specific amount of propellant we need for our operations will vary from the minor course corrections of proto-ballistae grabbing debris to the return of ten thousand tons of high-grade ore from Eros to high Earth orbit. Whatever specific amount of propellant we need, the celestial smelter can make it for us. The propellant stations will simply be a source of free fuel to lower the use of the smelter.

As Space Corporation grows, we will build more propellant stations to have a cheap propellant alternative to nuclear power. But since this process is so similar, we will always be able to use our smelters for both ore smelting and hydrogen propellant production.

In an emergency, our equipment could even use water and oxygen as propellant, since the extra surface reactor will be able to run hot enough to split water for use as propellant. This is what the celestial propellant package does when connected to an extra solar reactor and a nuclear thermal rocket.

For the first ballista mission in year five, we will launch the ballista with a full liquid hydrogen tank. This will get it to the ice asteroid, at which point the CPP will deploy and permit the reactor to melt small parts of the asteroid itself while pushing it toward Earth. Hydrogen will really only be necessary when we are moving ore, since we will not be able to efficiently melt iron into a gas and fire it out of the rocket as exhaust. I have seen metal get vaporized, and while it looks really cool, it is expensive, dangerous, and energy inefficient. Melting ice is a lot easier.

Our movement to the stars will only be possible if fuel is waiting for us in space. Our first space infrastructure will be made from things already there. Table 3 lists large derelict objects in Earth orbit.

Table 3. Derelict objects in Earth orbit

Name	Mass	Name	Mass	Name	Mass
Alouette 1	145	Arabsat 2A	2,500	Kwangmyŏngsŏng-3 Unit 2	100
Alouette 2	146	Arabsat 2B	1,570	NEXTSat	224
Anik A1	565	Astra 1A	1,780	Olympus-1	2,595
Anik A2	1,238	Astra 1B	1,562	Ratsat	165
Anik A3	286	Astra 1C	1,700	Rossi X-ray Timing Explorer	3,200
Anik B1	887.2	ASTRO	700	SNAP-10A	290
Anik C1	563	ATS-3	365	TDRS-1	2,268
Anik C2	563	Cerise	50	TDRS-4	3,180
Anik C3	563	DART	360	Turksat 1B	1,743
Anik D1	633	Envisat	8,211	Turksat 1C	1,062
Anik D2	633	Injun	16	USA-165	100
Anik E-1	1,781	INSAT- 2DT	1,360	Vanguard 1	1.47
Anik E2	1,781	Intelsat 1	34.5	Arabsat 1C	600
Arabsat 1A	600	Intelsat 601	4,259.2		
Arabsat 1B	600	Kosmos 1805	1,400		
Total					53,330 kg

GREEN STATION 1

During the Apollo program, all of the rockets were designed to bring the fuel tank from their third stage up with them. These tanks were all drained dry by the engine's push into space, but the empty shells are still in space. These tanks don't orbit around Earth, but they orbit around the sun (roughly in the vicinity of Earth). The dimensions for each fuel tank are 17.8 meters in height and 6.6 meters in diameter. One fuel tank is 986 square meters of internal surface area. If we attach all four of them, with a hole of two square meters in the center to connect all of the sections, we will have a very cramped 3,936 square meters of surface area.

On the outside, all four of these sections will be connected to share shielding, power, and navigation control. The entire object will be a cylinder 71.2 meters long and 6.6 meters in diameter.

The upper stage for *Apollo 12* is also thought to have been discovered and falsely labeled as an asteroid (J002E3). If we found all five of them, that would be even more space.

One fuel tank of this size was used to build the Skylab space station, which fit a three-man crew for several months, plus all of their experimental equipment and living area. No humans will permanently live on Green Station 1. They will visit for a few hours, plant crops, harvest crops, and then leave. To collect these old fuel tanks, we will wait until they come near Earth and then slowly collect them with proto-ballistae and ballistae.

Once the celestial smelter is launched in year six, we will have an orbital scrapyard, which at a minimum will include four Apollo fuel tanks, Envisat, Intelsat, the Rossi X-ray Timing Explorer, and Olympus 1 (see table 3). We could probably have every single piece of space debris on my list in that time (even the tiny Vanguard), but we'll start with these seven objects.

Each upper stage will have its engines removed from the fuel tanks. These engines are made of very valuable, high-temperature metal that will be useful for making ballista-1 or 2s in the future. The fuel tanks themselves will initially be strapped together by proto-ballista. All of the solar panels and electrical wiring from the four large satellites will be removed by the proto-ballista. If possible, aluminum components will also be removed from these satellites.

The smelter will use the nonaluminum, nonsolar panel remainder of each satellite to create a weld pool (liquid metal that, when cooled, connects other pieces of metal into one piece). This will make four to five Apollo fuel tanks one large cylinder.

We will then remove the solar panels from all of these derelict space vehicles (which can make up as much as half of the mass) and attach them to the outside of Green Station 1. This will provide thermal insulation, radiation shielding, and a small amount of electrical power.

These panels are usually designed with power generation as the primary purpose. A green station will need very little power—its purpose is to make food cheaply. By wrapping the station in solar panels, we will optimize our time, mass, and money.

We will probably find some panels that are no longer operational. These can be placed under the first layer, since they are still good thermal insulation, and will help block radiation from space. If solar panels are not enough to buffer the temperature and radiation, we can add extra layers of melted debris. Plants can withstand much higher radiation levels than humans, so we will only provide as much radiation shielding as the plants need.

When Green Station 1 is finished, we will have a large amount of solar power left over. We will build a small hydrogen reactor that will use all excess electricity to split oxygen and hydrogen. This is the same concept as Propellant Station 1 but on a smaller scale.

Most derelict satellites, or used rocket stages, have their own fuel tanks. All we need to do to make hydrogen out of water is get an old fuel tank that can fit about 1,200 kilograms of water. We will place two cathodes into the tank and then run an electric current through it. This will separate the oxygen from the hydrogen. The cathodes can take up a small amount of space at the fuel tank's opening, and we can place a filtration system there to siphon off the hydrogen and oxygen. Propellant Station 1 will be the main depot for this, but we can also generate a little propellant from the excess electricity at Green Station 1. We will have to move ice to the green station anyway while we are growing the plants, so it will not be a big inconvenience.

Based on the number of solar cells we can recover from all of these derelict space vehicles, we may greatly extend the number of panels on Propellant Station 1. Other than the first few missions to develop the infrastructure in space, we don't plan on flying any propellant to orbit.

Whenever possible, parts of derelict space debris will be integrated into Space Corporation designs. There are dozens if not hundreds of rocket motors and fuel tanks of various sizes floating around Earth. We can use any of these fuel tanks to store water for hydrogen production, food growth, or oxygen storage. As our presence in space expands, we will mass-produce more standard components. Our first few cars will come from the junkyard.

We will launch our first ballista in year five, and it will not return with an ice asteroid from the main belt until year seven. By then, Green Station 1 and Propellant Station 1 will be finished; they'll only need water to start generating hydrogen propellant, oxygen, and plants.

We may fly up a small amount of water to do experiments in Green Station 1 while we are waiting. There have already been plants grown in space, but this was not for large-scale human consumption.

Once the work is done on the outside of Green Station 1, we will add an airlock, communication equipment, and a power interface on one end. We will simply cut out a section to be used as an airlock and attach a few remaining pieces to the outside.

Before the interior is pressurized, we will fly up four thousand specially designed, plastic hydroponic containers and anchor them into the interior walls of Green Station 1. We will remove pieces of ice from our asteroid in orbit, fill each hydroponic container, and move them inside. We will not pressurize the green station until all of the ice and hydroponic bags are moved in and attached. Eventually, when the station is heated, the ice will melt, and the seed bundles can be added through a small valve at the top. We'll design the bags to have a few hard pieces of plastic on the inside for plant roots to anchor to.

We will put ultraviolet lights on the inside of each fuel tank. These can be timed to give a comparable day/night cycle to what the plants are used to on earth.

All of the actual construction work for these craft can be done by proto-ballista, with humans controlling them from Earth. Even tending to the plants on Green Station 1 can be done with remotely controlled variations of NASA's existing Robonaut (I refer to the Space Corporation version of this as a worker robot).

It would not be hard to make a small bundle with seeds and nutrients and fly four thousand of the bundles on every manned flight. Depending on the species 4,000 seeds could weigh a few kilograms. For many species the seeds are the edible part of the plant (potatoes, nuts and beans). We can probably keep one 20th of the crop for seed and eat the rest.

We will time crop cycles for three-month intervals to correspond with mining expeditions or Mars landings. Every three months, a crew flies up, harvests the crops, and then replants the seeds.

We will make it a point to collect CO_2 from our own crew and the space tourism business. As this is a waste product, it should not be expensive, though we'll have to standardize the use of our air batteries throughout space. (Money helps with consensus building.) We'll do the same for any human or plant waste to make food for the plants. The oxygen by-product from the propellant station will also be an easy source of oxygen.

The basic idea of doing crop rotations on Earth can be applied in space, with one-third of the hydroponic containers growing certain crops and then all of them being rotated per each harvest. We'll experiment with several different species. Some are good for oxygen production, others for food. Nutrition and ease of cultivation will also be factored in.

Green Station 1 will at a minimum be able to produce three months of food for a human crew of four at a state of equilibrium. It's likely we can grow way more than that, but that's the base amount we'll factor into the plan.

As each crew docks, they will remove three months of air from Green Station 1. Just before the crew returns to Earth, they will attach their CO_2 tanks to the green stations so the plants can breathe. They will also leave any waste products from themselves or from the plants. If we grow more food than the crew needs, we'll sell it to the space tourism business. We can also freeze-dry some of this food so that it will be available as the space business expands or as emergency rations.

While both the propellant station and the green station will specialize in their respective spheres, there will be a lot of cross-functionality between them. The green station will have some excess electricity since it is covered in solar cells, so a small amount of hydrogen propellant can be made there (attached externally). Propellant Station 1 will have internal areas that are vacant, and these can be converted for use as a smaller green station.

As time goes on, we will build more and larger propellant and green stations from space debris and nonmarketable asteroid metal. Mars, Venus, and Ceres will also need these stations, since that's where we will be working. Eventually we will make huge cylindrical propellant and green stations, but it will be a few years until demand gets to that level.

Once we have utilized all of the large debris in Earth orbit, we will round up debris in heliocentric orbit. This is debris that orbits the sun, not Earth. We will use proto-ballista for maneuvers in Earth orbit and then use the ballista to move objects out of heliocentric orbit. The following table is a list of debris orbiting the sun near Earth's orbit.

Table 4. Derelict objects in heliocentric orbit

Name	Mass	Name	Mass	Name	Mass
Contour	775	Mars 7	2,000	Stardust	300
Genesis	494	Mars Observer	1,018	Ulysses	370
Helios	370	Pioneer 4	6	Vega 1	4900
Herschel Obs	3300	Pioneer 5	43	Vega 2	4900
Luna 1	361	Pioneer 6	146	Venera 1	643
Mariner 10	474	Pioneer 7	138	Venera 2	643
Mariner 2	202	Pioneer 8	146	Venera 11	4,940
Mariner 3	260	Pioneer 9	147	Venera 12	4,940
Mariner 4	260	Ranger 3	329	Venera 13	760
Mariner 5	244	Ranger 5	342	Venera 14	760
Mariner 6	411	Apollo 8	9,500	Wikinson	840
Mariner 7	411	Apollo 9	9,500	Zond 2	960
Mars 1	893	Apollo 10	9,500	Zond 3	960
Mars 4	2,265	Apollo 11	9,500	Phobos 1	6,220
Mars 6	1,370	Apollo 12	9,500		
Total					**86,541**

TINY SPACE JUNK

We are not cleaning up debris in Earth orbit because we really, really hate littering. We are doing it to procure cheap scrap metal at the top of the gravity well for profitable purposes. The least profitable pieces of space debris are millions of tiny pieces, and these will probably be the last pieces we clean up.

Space Corporation has a bigger interest in cleaning this mess up than anyone else. (It is literally in the way of us making money.) As space commerce increases, the risk of collisions with floating debris also increases. Plus, the space vehicles of the future will not be government science projects with no value on the free market. If a two-hundred-kilogram chunk of gold were knocked into the Pacific Ocean by a one-hundred-gram piece of an old Soviet rocket, we might have to reduce our employees Christmas bonus by half a percent.

While the proto-ballistae and even the ballistae between mining expeditions will be able to grab large debris and fly it to orbital scrapyards, we will want to invent a completely separate system for tiny debris. Picking up (pardon the spatially incorrect phrase) something the size of a potato chip is very different from grabbing something the size of a fat dog. After the first nineteen thousand reasonably large pieces have been accumulated with mechanical arms, we can fit the proto-ballistae with extra attachments or nets.

The Japanese *Hayabusa* mission flew behind a comet and used gel capsules to capture a large number of tiny pieces from the comet's tail. This same system can be adapted for our purposes. Hopefully Japan will sign the Space Corporation charter, and the same team of scientists can adapt the *Hayabusa* device to be held by a proto-ballista.

We will have to add a way to compress the fragments out of the gel and into a storage vessel, optimizing the amount of debris one gel pack could collect. We will still keep all of these potato chips in space. There are hundreds of tons of debris in these small pieces, and many of them are still made out of useful metals (aluminum, titanium, etc.) It is even easier to melt tiny pieces than it is to melt large ones (they fit into the smelter better).

The primary goal for Space Corporation is for-profit asteroid mining, but there will be delays between mining expeditions due to orbital alignments. Whenever the equipment is not being used for mining, we can control it from the ground to go after any space debris nearby. Someone has to shovel the snow from Earth's driveway.

Even if this debris is not used in the first phase of colonization, we are doing this for the long term, and we may find a use for it someday. Indeed, there is no reason to think that once we amass all the space junk we can't *then* design things out of the resources we have instead of lamenting the expense of our current designs.

When we start to get mass from asteroids instead of from junk, the knowledge and equipment of smelting in space will be retained. As the efforts of the corporation progress, more and more components will be made offworld. Eventually, all we'll need from Earth is wireless Internet and customers.

MEGA-STATIONS

Now that I've explained the concepts of green stations and propellant stations, I will describe their large-scale cousins, mega-green stations and mega-propellant stations.

As the space-mining industry expands to cover a larger quantity of metal, more abundant, nonmarketable metals will be available as the by-products of space mining. We can begin to construct massive green stations with these materials (enough to plant one thousand trees or ten thousand small plants).

To bring hundreds of thousands of tons of metal to Earth each year, we'll need more propellant than can be generated from solar power. We will make mega-propellant stations, set to the same size as a mega-green station but with a few minor differences.

Scientists have observed since the 1950s that space stations designed as large wheels can create artificial gravity from their own rotations. A station like this would have to be built on a large scale (not useful for mining expeditions or welded space junk). But, once the rare earth metal market is being moved through Space Corporation smelters, we'll have more than enough unmarketable metal to start building these facilities.

This change in size will also be useful for our change in activity. The small green stations and small propellant stations are a good start to lower the cost for small four-man crews making three-month trips between Earth and Mars. This will be enough to start bringing in billions of dollars of metal. Within a few years, this will dominate Earth's mining market. We'll start to bring in trillions of dollars, and there will be thousands of people in space.

With every kilogram of marketable metal, there will be at least twenty kilograms of unmarketable metal. We will plan from the beginning to make standard hull plates so as to design a large wheel-shaped station that is 400 meters in radius, 2,500 meters in circumference, and 300 meters in height. The center part of this wheel will be empty. This design will be used both for mega-green stations and mega-propellant stations.

This size was selected because it is big enough to fit one thousand tipuana tipu trees. I will explain why I selected this tree during chapters 7 and 8. Of course, at first we will probably be using all of this space for smaller food crops, but we want to use the same structure for growing anything from trees thirty meters tall to tiny alfalfa plants.

We can layer several sections for smaller plants as we get closer to the interior of the mega-green station. (Smaller plants will not be as affected by gravity as large trees.) We can set up a system of mirrors so that each mega-green station has multiple levels on which to grow crops. We can even bring herbivorous animals up to these stations, but this would be more for research than viable food production.

As each pili is melted down, all of the nickel-iron will be used to make standard hull plates. In this way, as ore is melted down and the valuable material is made into an ore bundle, the unmarketable metal will be put into a plate that is slightly curved and has interlocking joints to connect with other plates. All of them will fit together very easily. Once the object is complete, we will melt debris along the small gaps where the plates are attached, making a sealed cylindrical shape. All we need to do is rotate this cylinder at the correct speed to simulate Earthlike gravity.

To reassure everyone on Earth, the material used to attach the iron plates will have a melting point lower than Earth's reentry temperature. We can also cast these plates very thin, so that they will melt when entering Earth's atmosphere. Even in the event of an accident, or a terrorist attack where one of these stations were to fall to Earth, all of the hull plates would separate, and they would individually burn up on their way down. Inside of the thin metallic hull will be a thick layer of water. This water is for tree roots on a mega green station, or for water to be split in a mega propellant station. This will be a large portion of the mass and will be distributed as water vapor if it entered the atmosphere from space.

We will coat the megagreen station with solar panels on the side that faces the sun. This will serve as a thermal insulator and will generate electricity for the station. There will be a small human crew on the megagreen station, just to study and cultivate all of the plants.

Just inside of the metal hull is where we will keep a small lake of water. This is what the crops, and even trees, will grow on. If we have to use soil, we can put nutrients into small pieces of pulverized material from mining. Yet another option is putting some rudimentary bars and pipes through the water reservoir so that the trees have something to anchor their roots to.

Besides being essential for plant life, this will help to buffer all of the people and plants from radiation. The dock to access the mega-green station will be in the center of the circle, where there is zero gravity. (This will also be the recreation area for the people that live there, tourists, and visitors.) Except for lighting and electricity, the mega-green station will be well planned slag.

The mega-propellant stations will be almost identical to the mega-green stations. The only difference is that the reservoir of water will have nuclear reactors running electricity through the water and producing hydrogen and oxygen. We'll use the same standard storage tanks to collect the by-products. Since hydrogen and oxygen are both lighter than water, they will naturally move toward the top (the center, which is the point of weakest gravity). The tanks will attach to the outside of the propellant station and will use molecular sieves to only let in hydrogen or oxygen.

This is the same general concept as mentioned for Propellant Station 1. Since we are only using gravity to split them up, we will have to layer our propellant and air containers. All of the hydrogen and oxygen will first pass through an oxygen sieve. Afterward, the hydrogen will pass through a second sieve and go into the hydrogen container. Once these containers are full, they will be detached and added to their respective vehicles. We can also use magnetic charges to separate them. (Hydrogen is positively charged, and oxygen is negatively charged.)

By standardizing our propellant and air production across Space Corporation as well as within the space tourism industry, we will be able to lower the per-unit cost of each item.

As human settlement in space continues to expand, Space Corporation can design mega-stations for use as offices or apartments. When billions of tons of space-mined metal are coming to Earth, there will be thousands of workers that stay in space, even when robots are doing most of the work. There will also be hundreds of humans going to or from Earth orbit, as well as tourists floating in space. With virtually no gravity or friction, these stations could be propelled by NSWR for movements to the outer solar system or for permanent living areas orbiting other planets. Eros 433 will likely be our first spaceport/mining colony. Since we're making these for profit, the possibilities are endless.

DEBRIS-RECYCLING CONCLUSION

Instead of leaving heaps of dangerous debris to collide and make more dangerous debris, Space Corporation will use these materials as the feedstock for our expansion into space. It may be that half a dozen problems, which cumulatively cost trillions to solve, could all be combined into one concept and resolved profitably.

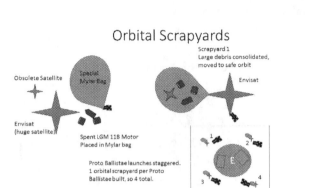

Image 2

Orbital Scrapyards

Obsolete Satellite

Special Mylar Bag

Scrapyard 1
Large debris consolidated,
moved to safe orbit

Envisat

Envisat
(huge satellite)

Spent LGM 118 Motor
Placed in Mylar bag

Proto Ballistae launches staggered.
1 orbital scrapyard per Proto
Ballistae built, so 4 total.

Image 3

Space Infrastructure

5 Apollo tanks welded together by Celestial Smelter. Smaller fuel tanks and solar cells removed from other extant space debris.

Salvaged solar cells coat the outside, providing some power. Food grown on inside. Water brought inside from ice asteroids. Small amount of water splitting from solar cells.

Celestial smelter can split water to make hydrogen propellant and oxygen.

Propellant Station 1 with additional solar panels taken from derelict space debris.

CHAPTER 5

A Profitable Space

The reason that no lasting human presence has been established in space is that doing so profitably is illegal. It is impossible to privately own any objects that are in space (unless they originated on Earth, like satellites). Certainly governments have waded into the oceans of space, and at enormous expense. The International Space Station has cost US$150 billion dollars with no financial return on investment. Humanity would have to go without food for a year to put an international hot dog stand on Mars. Actually I'm breaking my own rule here. Mankind did not pay $150 billion dollars for the space station. The cost was not very international, 4% of humankind (I'll let you guess which 4%) paid for two thirds of the station and flew 36 missions with their space shuttles to assemble it. If Space Corporation were funded this way America would own almost the entire solar system and Canada, Russia, Japan, and ESA could all share Titan. I've heard it's really nice this part of the plutonian year.

How many meals are provided by the Christian Children's Fund versus the amount provided by McDonalds? Certainly McDonalds has given millions of people jobs and done very well for its stockholders and its customers. I do not believe three separate moral actions make McDonalds immoral. I'm not interested in mankind slowly crawling to space on the throats of millions or billions of taxpayers. We can go to space as quickly as we adopted the use of electricity, airplanes, steam, or steel. There is no charity where we are going.

I briefly mentioned my plan for colonizing space to my brother and he asked me "yeah that sounds great, but what will this cost the government?" This got me thinking. A huge part of the US population has iphones. How much did the US government pay for these iphones? The answer (during 2014): negative 14 billion dollars. The United States government didn't pay a penny for iphones. Apple Corporation created them, manufactured them, shipped them, convinced people to buy them, serviced them, upgraded them, and created most of the apps and software for them. Apple Corporation is very grateful that the US government hasn't stolen the iphone and made it the collective possession of mankind, so they gave the US government 14 billion dollars in taxes, in a single year. If the US government was as amicable toward space corporation they could make 14 trillion dollars in tax revenue, forget 14 billion.

My plan is easier to implement if we receive equipment and technology from the various democratic governments of this world. We will still want large amounts of private capital raised from businesses and individual people. I don't have a crazy idea for this; rather, it is actually a four-hundred-year-old idea—a joint stock company. We will issue shares in future profits to investors. The start-up capital will be much lower than any national space agency because none of the technology in my plan is new.

Imagine if someone wanted an ice-cream machine attached to a Jet Ski. It is very likely that such a configuration has never existed. Nonetheless, there is nothing physically impossible about it; indeed, thousands or even millions of both of these machines exist all over Earth. If both a Jet Ski and an ice-cream machine could be provided in exchange for stock (a share in future profits), we'd only need one hundred dollars for a welder and a plumber to connect them.

Some advertising would be geared toward science enthusiasts: "Do you want to buy shares in the universe?" For people who are fiscally oriented, pointing out that gold passed US$1,900 per troy ounce in 2011 could bring many of them along. Space Corporation shares will yield a dividend based on profitable return from the space-mining business. There are already several space-mining businesses. Some of these companies openly mention legal challenges ahead of them, since they plan to come back with thousands of tons of gold and pray no one sues them or steals it.

Without a legal framework for colonizing space, we might bring a few nuggets of gold back without a legal challenge. But after space commerce intensifies, there will be no way of establishing a legal claim to asteroids. Would anyone on Earth build a mine if profiting from mines was illegal?

When NASA landed a probe on Eros 433 in 2001, a man named Greg Nimitz filed a claim for the asteroid and then sent NASA a bill for parking their probe. This was dismissed in court because Nimitz was unable to prove actual ownership rights. While this story seems amusing, it begs a huge problem: if there is no way to legally own property in space, no one will ever develop space real estate. Mr. Nimitz has not placed parking meters, a hot dog stand, or a low-gravity water park on Eros since this court ruling, and I don't blame him.

The government lacks the physical means to police claims in space or on Earth concerning space property. If Pepsi is delivering sodas to Walmart and their truck crashes, we have a law called the bill of lading, which deals with the delivery of goods (who is responsible for it, how to claim damage, etc.). Not only do we not have laws like this for space business, but the Outer Space Treaty prevents us from even making them.

If any individual, company or nation built a lemonade stand on an asteroid any other individual, company or nation could steal or destroy it without breaking the Outer Space Treaty. The Treaty doesn't ban force, it only bans private property.

A few brave and wealthy souls could go to space anyway and hazard the challenges later, but the legal vacuum will discourage a large proportion of investment. Cars, airplanes, steam engines, and electricity were all developed at a time when the pioneers of these industries had complete proprietary control of their inventions, industrial plants, and finished products. If Thomas Edison were not able to legally make a fortune by opening electricity to man, I doubt he would have spent millions of dollars on research to figure it out. Johannes Guttenberg took out a loan from a bank to create the printing press. Apple spent 2.6 billion dollars to create iphones. Making something unprofitable has basically the same effect as making it illegal.

Subsidiary real estate development companies for Mars and Venus will also provide investment vehicles, since their value will go up as our real estate development venture gets closer to completion. Expectation of future success will drive the stock price, just like every other successful business for the last four hundred years.

Since there are already billions of dollars in civilian space businesses—any or all of which would gain from standardizing their power, fuel, launch rockets, or survey data with those of Space Corporation—funding this endeavor will not be difficult.

At present, the only legal way to fund space colonization is to beg for spare change on the steps of congress. Even then, the International Space Station is set to be abandoned in 2024. We are also planning to spend 30 billion dollars so two astronauts can high five on the surface of Mars.

Dividends from the stock would be paid out after the first pieces of a metal-rich asteroid is smelted and brought back to Earth (around year eight). Since only free people and governments will be allowed to participate, we will be spared the "blood for oil," "blood diamond," "blood for everything made by China" arguments going on now. Public knowledge of the company will go out just like any filing with the US Securities and Exchange Commission. Success will bolster investment, and failures will lower it. Philosophy and physics don't change only technology and opportunity.

ICE RETRIEVAL

In chapter 4, I explained how space debris could be built into propellant-producing and fuel-producing infrastructure. Before we can use water to make propellant, food, and air, we have to bring it to Earth orbit. We can do this with tiny atomic devices. Specifically, we will take the W54 atomic weapon that the US military made during the Cold War and slightly modify it for use in space. This device weighed only 23 kilograms (51 pounds). Nuclear physics is such that this tiny machine was able to produce the same force as 250,000 kilograms of TNT. The modified version used by Space Corporation will be called the "small atomic mining device."

In year four, we will launch the first Katas to go toward the Kuiper Belt. We'll have the Katas scan the main asteroid belt to determine the closest ice asteroid to Earth with a mass of approximately twenty thousand metric tons. The Katas will mark this ice asteroid with a radio tag while en route to the Kuiper Belt.

In year five when we launch a ballista, we will direct it toward the designated ice asteroid. The ballista will be carrying a small atomic mining device. This device will be very closely guarded by the contributing state until it is launched and then tracked by the United States and Russia after launch.

Due to the warmth of the sun, ice evaporates in direct sunlight anywhere sunward of Mars. It takes a nuclear thermal rocket three months to reach Mars at its closest orbit; I'm assuming it will take an additional three months (so six months one way) to reach our first ice asteroid from Earth.

The ballistae will be equipped with MA-2s, large mechanical arms that can be heated so that they can attach to ice objects. We will slightly modify this set of MA-2s so that they can melt through a small amount of ice and use a CPP-2 to move the steam/water out of the tunnel before it refreezes. This will result in small tunnels being made in the ice that are the same diameter as the ballista's arms. Let's assume that the shape of this asteroid is a cube. Since one cubic meter of ice has a mass of 910 kilograms, we know the dimension of a 20,000-ton ice asteroid will be twenty-eight cubic meters. Practically speaking, the asteroid is more likely to be rectangular than cubicle (since its gravity will be too low to have made it a sphere). In either case, we'll assume it's about twenty meters thick at its narrowest point near the center of mass. We'll use the arms of the ballistae to melt ten tunnels from the exterior of the asteroid down to the center of mass, with all of the tunnels existing in the same geometric plane and intersecting at the same point. We will then have the ballistae place the small atomic device at the intersect point and fly to a safe distance.

We will time the device to go off based on the asteroid's natural rotation. When the geometric plain with tunnels is perpendicular to earth's location in 6 months, we will detonate the device.

The energy will follow the path of least resistance, splitting the asteroid along the sunward/plutoward axis. When the atomic device is detonated, the geometric plane with the tunnels will experience the most force. This will split the object in half just before the pieces of the object are thrown outward, resulting in one pili going sunward and another pili going plutoward. This is the basic idea behind Newtonian ballast, which I will use a lot throughout the rest of my book.

Anyway, since one pili is already moving toward Earth, it will be easy for the ballista to attach to it and make minor changes to its course to bring it into a stable high Earth orbit. The plutoward piece can keep floating until it leaves the solar system or is captured by one of the gas giants. There is no safety risk to people on Earth because an ice object of this size (10,000 tons) would melt before getting to Earth's surface.

It will take six months for the ballista to reach the ice meteor, a few days to cut tunnels into the asteroid, and then roughly another six months to guide the sunward piece to Earth orbit. By year seven at the latest, we will have our source of water orbiting the Earth.

PROFIT PROJECTIONS

It was difficult to find specific numbers for the total amount of marketable metals that are mined on Earth. Generally US$1 million per ton is the price of rare earth metals, when taken as an average. Marketable metals (many of which are rare earth metals) would be even more than this. The estimate for human demand for rare earth metals is 185,000 tons per year by 2015 (hence $185 billion). Gold and platinum are not rare earth metals, but they are even more expensive, with their global market values being around US$100 billion for gold and $10 billion for platinum. For the sake of simplicity, let's pretend there's no human population growth, no growth of wealth on Earth, and no inflation. This will still be a market of roughly US$300 billion per year in marketable metals. From the first mining expedition in year eight to the Mars impact event in year forty, the company can make US$9.6 trillion. This is not counting all of the money from space tourism, planetary cruises, and delivering fuel to low earth orbit.

ASTEROID MINING EXPLANATION

The phrases "space mining" and "asteroid mining" are frequently used to describe getting metal from asteroids. This is sadly a by-product of science-fiction writers beating businessmen with planning space development. A mine is a structure dug under the earth's surface. When children pick up rocks in a field, they are not called miners. On Earth, we have to mine things (dig under earth for them) because few of these valuable metals are on the surface, and what was naturally on our surface was picked thousands of years ago. (The Romans depleted many of the gold mines known to them by AD 300.)

It is now cost-effective to dig up one ton of dirt to get 0.1 ounce of gold. Kalgoorlie, Australia, has a vast open-pit mine—basically it looks like a giant crater, except it was dug out with mining equipment. When we mine an asteroid, we can directly scan its surface, grip the top of a cherson, have an asteroid torch melt or cut off as little as possible to separate it from the asteroid (turning the cherson into a pili), and then fly the broken cherson to a smelter. This isn't quite the "sweaty men with flashlights on their helmets using pickaxes" that mining usually denotes. To use a more accurate description, I would call it "asteroid clipping," or "asteroid utilization" (it's very possible we can carry rocks off of an asteroid and smelter them). I'll still call it asteroid mining since this is the way that everyone else refers to it. In the future, it is possible we may dig a "mine" on an asteroid, but we can probably make a few trillion dollars before that happens. In the long term we will remove ore from asteroids in the shape of a mine so that we can build living quarters inside them (free radiation shielding). After a few decades there will be entire cities inside of hollowed out asteroids. But this will happen after a large amount of money has been brought in.

The Katas will scan and look for small asteroids near Earth before they go further into the solar system. Our first mining expeditions will intercept small asteroids coming near Earth. Collecting a five meter wide asteroid is about as difficult as collecting a five meter wide satellite.

Our celestial smelter will probably have to be launched on a heavy-lift rocket. We'll design the smelter to be large enough for the first ten relatively small asteroids that we plan to mine. Anything larger than that, and we'll use an asteroid torch to cut them into workable pieces. Arc welding will be easier to do in space than on Earth, because there will be no concerns of air mixing with the cooled metal and ruining its purity. On Earth, complex pumps are used to flood the area of a weld with inert gases (like argon) and keep away reactive gases (like oxygen). Metallurgy will be easier in space than on Earth.

SMELTING

Here is a quick note on precisely how smelting works: Essentially, a naturally formed piece of rock is placed in a very high-quality oven. Let's say that we grab a pili that's a mixture of water ice, gold, and iridium. Ice melts at 1°C (assuming earthlike air pressure in the oven, which we can control). Gold melts at 1,064°C and iridium at 2,400°C. We can adjust the temperature of our smelter and then siphon off any material that has been liquefied.

At 10°C, the ice will liquefy, and the water can be sucked out using the vacuum of space. We can collect it all when it is let outside (by a second container opposite the hole where the substance is sucked into space), or we can just discard it. We would then turn the temperature up to 1,100°C. Whatever we melt off at this point is gold, and we can harvest it in a similar way. (It will be in liquid form.) We'll move the gold into another container and let it cool. (All solids are frozen liquids.) If the gold cools in a container shaped like a teardrop, it will be a teardrop-shaped solid. We can then raise the temperature to 2,500°C in order to melt the iridium that we have left over, or if we're confident in the composition, we can just leave it alone since everything that's left would be iridium. In this case, we would want the water stored by the crew and the gold and iridium built into separate ore bundles. (Assuming an ore bundle standard mass of two hundred kilograms, this is US$310 million for the gold ore-bundle and US$120 million for the iridium bundle.) It's likely the iridium would go to a different part of Earth than the gold, so we'd want them bundled separately.

The useful metals that are not worth exporting to Earth (nickel melts at 1,464°C and iron at 1,538°C) we will directly put into molds to make support beams or some other useful thing for space infrastructure. Instead of using a mold shaped like an ore bundle (teardrop), we'll use a mold shaped like a support beam or a hull plate. Slag is the word for material extracted by mining which is a low-value leftover. In space all of our slag will go to build solar cells and space infrastructure.

We can make a small opening in the celestial smelter and use the vacuum of space to move the molten metal around for us. If we get the spacing just right, the molten material will cool as it leaves the smelter, and by the time it hits the mold, it will freeze into a solid shape. We'll place a different container near the mouth of the smelter each time we raise the temperature, to keep all of the products separate.

The smelter can have a conveyor belt built around it, which moves a thin mold in front of the opening. The vacuum of space will provide the force needed to move the liquid out into space and into the mold. Since the interior of the CS-1 will be something with a high melting point, any metal touching it will be liquefied and then pulled into space by the vacuum. We will use this force like a can of spray-paint, directing the liquid metal out exactly where we want to. We can close this hole while moving a new mold into place (during which time all the liquid will float around the interior of the smelter).

On Earth, nuclear power is restricted, and other elements are abundant. Smelting metal frequently has many steps and uses many different materials. In space, this will not be necessary (heat is sufficient and abundant, thanks to nuclear power). Eventually we may be able to use more complex methods of smelting in space if they save us money (smelting in zero gravity will have many affects not present on Earth).

CELESTIAL SMELTER DESIGN

The smelter will be roughly spherical in shape but with a small nuclear reactor in the center of the sphere. The outer sphere and the inner sphere will both be made of tungsten (melting point 3422°C). We may find a cheaper alloy of other metals, but we'll say tungsten right now. We can pump in whatever gas we want for the mining process by keeping that gas as a frozen solid nearby and melting it inside the chamber.

All of the surrounding surfaces will be made out of tungsten, so they will have a very high melting point. There will be a part of this sphere that can be opened to space at various sizes. When we are placing satellites or asteroids inside, we'll open the reactor up to a large extent. When we are using the vacuum of space to siphon liquid metal out, we'll make that opening very narrow. None of the metal will accumulate on the tungsten surface because its temperature will be high enough to melt any solid metal to a liquid.

Let's say we start with a sphere that is thirty meters by thirty meters, with a small nuclear reactor (one meter by one meter) in the center. This will be a different design from reactors on Earth, but it will be fundamentally similar. Simply placing pellets of U235 together can emit a large amount of heat, since the nuclei being emitted strike other nuclei, which has a chain effect creating more heat.

This heat will be moved through the tungsten, into the chamber, where the ore will melt. We can use some inert gas to transfer the thermal energy and regulate the air pressure. Most of a nuclear reactor on Earth is designed to absorb heat, convert heat, or limit heat.

Imagine Godzilla is your loyal pet. He is tied down from head to toe in chains, but vast amounts of wealth can be made simply by his pinky moving things for you (on Earth). Once Godzilla is in space, we can let go of his leash. Then he turns into a two-thousand-ton, green money-making machine! Good job, boy!

In space, we have no pollution concerns and no one-hundred-year-old power grids to supply a tiny amount of power for. The human operators will be as far away as we want (even on Earth's surface).

We'll start by taking apart derelict satellites, removing wiring and solar panels, and then melting the rest down. As these satellites liquefy from the heat, they will be sucked out of the small opening into space and then into a mold for whatever shape we want. While building the green station, we'll directly spray the melted metal onto the outside to attach the sections together or coat them with desired elements.

Even putting Green Station 1 together can be done with this smelter. The small orifice for the smelter will be set opposite the joints in the connected fuel tanks. As the liquefied metal is sucked out to space, it will fly one meter and touch the cold metal right at the joint between the two fuel tanks. We can have a proto-ballista gently spin the green station around so that the entire outside of the cylinders between fuel tanks is coated and connected with metal. We can also use melted metal as a weld pool to attach solar panels to the outside of the station.

The heat of the liquid metal can be altered by increasing the distance from the smelter to the destination. A smelter spraying molten nickel from a distance of three meters will give the nickel only three seconds to cool. Spraying it from ten meters will give it ten seconds to cool, at which temperature it will not have the energy to melt through the aluminum but would merely solidify (freeze) on contact with it.

The solar panels and electrical wiring removed from other space debris can be put on Green Station 1 intact. Solar panels will generate what little electricity the station needs for lighting, communication, and temperature. They will also serve as a thermal insulator against sunlight. As the solar radiation strikes the panels, a lot of it is consumed by the thermoelectric effect. What energy is not converted to electrons is dissipated into space, since the silicon in solar cells is a bad thermal conductor. We will cover Green Station 1 with working solar cells so that whichever way it rotates or orbits, it will be buffered from the sun.

We will start with very small asteroids, so there won't be any cutting needed at all. An asteroid that's ten meters by ten meters would weigh probably thirty thousand tons. If it has the same composition as Eros (0.5 percent platinum), then that would be 150 tons of platinum. This is equal to all platinum mined on Earth in a year. We will also be left with 29,850 tons of nickel-iron and small amounts of other elements useful for space infrastructure.

ASTEROID MINING BY PHASE

Phase I: Unmanned Mining inside of the Moon's Orbit

Asteroids frequently come near Earth. These usually create a buzz in the media and unfortunately a buzz over the heads of people in the Russian town of Chebarkul (as in February 2013).

In year four, we will have proto-ballista rounding up debris in Earth orbit. This will help us assess the design and software that we are using to guide these machines. As the Katas start to be launched, each one will conduct a gravimetric scan of a near-Earth object while looking for other smaller objects in the infrared spectrum. In year six, we will launch our celestial smelter, which will give us the ability to melt down objects in orbit. We will get the smelter started working on derelict space debris. Once a small meteor comes near to Earth, we can divert a proto-ballista to push this asteroid into a stable orbit and feed it to the smelter. We will also launch a few ballistae around this time, which are much stronger and faster than proto-ballistae. The ballistae will allow us to intercept a much larger number of passing asteroids. These first missions will generate a few hundred million dollars, boost the confidence of our investors, and help us improve our systems, software, and training.

I will use a less dramatic example than Chebarkul. In 2002, the asteroid 2002 MN came within 123,000 kilometers of Earth. It was sixty meters in diameter. Since a nuclear thermal rocket travels at eighty thousand kilometers an hour, were all of the space equipment in orbit already collecting orbital scrap, we could have sent machines to intercept this object in two hours. We could have spent a few hours, even a few minutes, removing chersons from the asteroid with the asteroid torch. We could then bring these chersons to earth orbit to be melted down at our leisure. Besides flying the equipment up, the only limiting factor is the availability of hydrogen propellant, and this will be available from our space infrastructure starting in year seven.

All of this activity will be controlled from Earth and must occur within a few hundred thousand kilometers of Earth (so the delay from signals at the speed of light doesn't bother the human operators). The Predator drones used when I was in Afghanistan primarily flew themselves; only a small amount of human interaction was required for certain critical steps. That is the level of expertise we want with our robotic space fleet.

It is much cheaper to control and operate these robots from the ground. We humans are delicate little snowflakes when it comes to industrial, aerospace, or nuclear engineering. (I melt at a much lower point than aluminum.) Controlling unmanned vehicles from the ground is not the way of the future; it's a starting point. This will allow us to conserve resources. We must use risk, money and personnel where they are most needed, deep space.

Ultimately, I don't want to make a few hundred million dollars. I want to shift trillions of dollars from inside of Earth's economy to the economy of the solar system. To do that, we are going to have to venture farther out into space and mine very large asteroids.

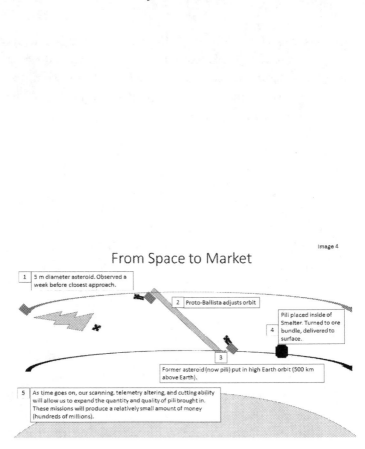

From Space to Market

1 | 5 m diameter asteroid. Observed a week before closest approach.

2 | Proto-Ballista adjusts orbit

4 | Pili placed inside of Smelter. Turned to ore bundle, delivered to surface.

3 | Former asteroid (now pili) put in high Earth orbit (500 km above Earth).

5 | As time goes on, our scanning, telemetry altering, and cutting ability will allow us to expand the quantity and quality of pili brought in. These missions will produce a relatively small amount of money (hundreds of millions).

Phase II: Manned Mining past the Moon's Orbit

In year four, the first Kata will fly through the main asteroid belt and find the closest ice asteroid with a mass of ten thousand metric tons. In year five, we will launch a ballista to go and bring back this ice asteroid. In year seven, that ballista will return to Earth orbit with several thousand tons of ice.

By the time the first ballista returns, Propellant Station 1 and Green Station 1 will be completed. I'm assuming a year to make our propellant and food in space. This is the year when short-range missions (phase I) will be going on, allowing us to gain expertise with space mining.

In year 7 we will send a ballista to divert several small ice asteroids into orbit around Eros. We will confirm that at least one of them is in orbit before we leave. We want water waiting for us at our destination to save money. We can use the smelter to break down the water for propellant and oxygen. We will need more propellant to get back because we will be coming back with ore. We can also use the water for food, drinking, and to produce breathable oxygen.

In year eight, we will launch our first long-range, manned mining expedition to bring back thousands of tons of material. The goal is to pay off everyone's initial investment in the first few years of full scale asteroid mining. Depending on how pure the metal is on Eros 433 a few thousand tons of smelted marketable metal could meet the 300 Billion dollar demand for metals on earth for a year.

Having fuel, food, water, air and space-mining experience will cut the cost of this expedition from billions of dollars to millions of dollars. How expensive would it be to carry a train on the back of a donkey? Let's just have some patience, build the train first, and use the train to carry the donkey. We'll also be confident that all of our technology works, and our investors will feel comfortable in us diving deeper into our galaxy of profits.

The first mining expedition will be launched from Earth in year eight. It will consist of:

- four proto-ballistae (pulled off space-debris duty)
- one habitation module with a four-person crew
- two ballistae
- two asteroid torches (launched specifically for this)
- celestial smelter

I use Eros 433 as my example asteroid. Of course the actual asteroid for our first manned expedition will be based on specific data that we get from the Katas and the rare earth metal commodity price index. We'll have to see.

Since all of our equipment can be interconnected, we will fly all of it there while attached to one ballista. This is the same concept as a railroad. One car is the actual engine, and the rest are simply towed as freight. The coefficient of friction is virtually zero in space, so moving all of these objects will require one small burn from one ballista and then another small burn later to slow down.

At its closest approach, Eros comes about fifteen million kilometers away from Earth. This is a twenty-day trip at the closest point by NTRO. We'll factor twenty-five days to get there a little early (just before it reaches closest orbit). We'll work on Eros for forty days and then fly several tons of ore back to Earth. Eventually (maybe even the first mission) we will permanently station a human team on Eros. It depends on when we get Green station 2 and propellant station 2 operational (in Erotian orbit). The ore can be set to coast back to earth by unmanned ballistae. The human crews will rotate as Eros nears earth each year.

Our Katas will have already flown by and given us detailed information on which chersons are the most valuable and easiest to remove. Some may honestly be boulders left from a previous impact, and we may not need to burn them out at all. The gravity on Eros is so weak that we can simply fly our equipment right up to it and then fly off again; we'll never have to land. Let's estimate that each day we'll remove one hundred chersons that are each ten tons from the surface of Eros. We will smelter each one on the spot, removing the marketable metal and making hull plates or support beams out of the unmarketable metal. Unmarketable metal will stay on Eros or in Erotian orbit. We will use these to build space infrastructure, ballistas, Hab shields, whatever will reduce our operating costs in space. The marketable metal will be put into ore bundles and attached to each other in series (like train cars). After forty days, we will have four thousand ore bundles. When we're done, we'll attach the ore to the back of all this equipment and fly back to Earth orbit.

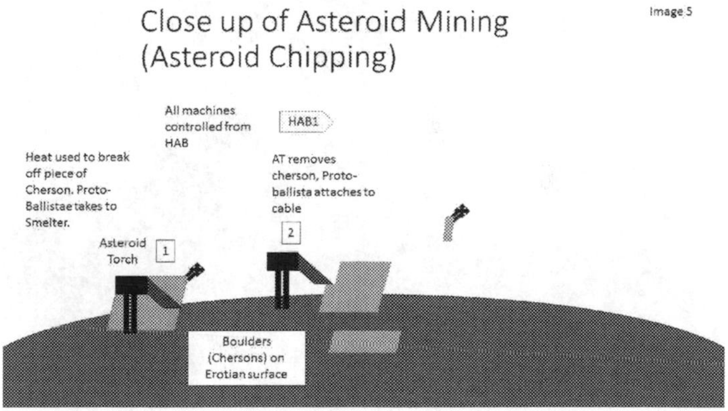

Close up of Asteroid Mining (Asteroid Chipping)

Image 5

All machines controlled from HAB

HAB1

Heat used to break off piece of Cherson. Proto-Ballistae takes to Smelter.

Asteroid Torch 1

AT removes cherson, Proto-ballista attaches to cable

2

Boulders (Chersons) on Erotian surface

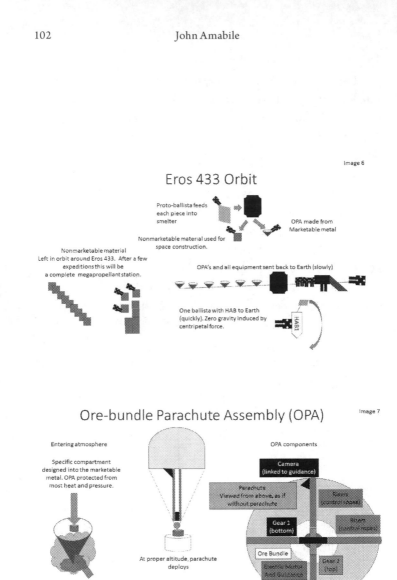

Image 6

Eros 433 Orbit

Proto-ballista feeds
each piece into
smelter

OPA made from
Marketable metal

Nonmarketable material used for
space construction.

Nonmarketable material
Left in orbit around Eros 433. After a few
expeditions this will be
a complete megapropellant station.

OPA's and all equipment sent back to Earth (slowly)

One ballista with HAB to Earth
(quickly). Zero gravity induced by
centripetal force.

HAB1

Image 7

Ore-bundle Parachute Assembly (OPA)

Entering atmosphere

Specific compartment
designed into the marketable
metal. OPA protected from
most heat and pressure.

OPA components

Camera
(linked to guidance)

Parachute
Viewed from above, as if
without parachute

Risers
(control ropes)

Gear 1
(bottom)

Risers
(control ropes)

Ore Bundle

Gear 2
(top)

Electric Motor
And Guidance

At proper altitude, parachute
deploys

Ablative Material
(melted off during atmospheric entry)

When the first human crew returns they will leave the HAB in space, board an Earth SSTO, and return to Earth's surface. They will have a hero's welcome and a huge pay bonus.

The ore will be sent on a safe high orbit from Eros to Earth. Whether the first crew stays on Eros for a year or rushes back in 2 weeks, the ore will probably not arrive at the same time the crew does. When the ore arrives near Earth OPAs will be added by proto-ballista. They will then move the ore bundles into specific flight paths to get them to our customers on Earth. As the Ore bundles enter the atmosphere the ablative portion (cheap metal or ice) will melt off, leaving the valuable portion and the parachute assembly. We have a lot of experience with ablative space material since Intercontinental ballistic missiles also use ablative shielding to reenter the Earth's atmosphere. Once the valuable metal is at a safe altitude the parachute deploys and the camera/GPS system will adjust the risers to bring the bundle down to whoever won the bid on eterraform.com.

Four thousand tons of pili, assuming that 1 percent of their mass is marketable metal, can be smelted into forty tons of ore bundles. At two hundred kilograms of marketable metal, that is two hundred individual bundles. At the price of iridium that would be $51 Billion. We'll ensure a good variety of metals so we don't crash the earth's market. The law of supply and demand would make a million tons of platinum worthless. We will bring back a few tons of as many different elements as we can. We can specifically mine 20 different parts of Eros to bring this about. One bundle will be for one element, so hopefully we'll have 20 of iridium, 30 for platinum 5 for lutetium, etc. Each ore bundle will weigh 300 kg when it is made. They are designed for 100 kg of inexpensive metal to melt off during reentry, but for the 200 kg of marketable metal to be intact when the parachute deploys.

Below is a specific list of marketable metals that are expensive enough to collect in space. Many of these metals are rare earth metals. I use the phrase "marketable metal" because gold and platinum are on my list but are not rare earth metals. Also, some rare Earth metals (like neodymium) are not expensive enough to make my list. Finally I call them marketable because it is based on the changing economy. If better wind technology is developed, the price of neodymium will go up, and neodymium may be added to the list. If a cheaper alternative to iridium were found on earth, its price would drop and it may come off the list.

Table 5. Marketable metal list

Elements	USD/kg
Americium	$1,500,000
Europium	$1,350,000
Lutetium	$340,000
Actinium	$100,000
Thulium	$70,000
Gold	$59,840
Rhodium	$56,000
Platinum	$47,000
Iridium	$42,000
Palladium	$21,798
Ytterbium	$14,000

Space Corporation will keep radioactive elements in space. If the level of radiation is very small, we will still import elements to Earth. (All matter is technically radioactive. We'll set the level to be low enough that a bad entry into the Earth's atmosphere will have no radiation for the Earth's surface.) Since most Space Corporation equipment is designed around nuclear technology, we will eventually need all we can get as fuel. We also don't want to take the unnecessary risk of radioactive material crashing into someone's property from space. If a two-hundred-kilogram chunk of gold accidentally crashed on someone's property, far from being sued, I doubt we'd even be notified.

We will want to find a large supply of uranium or thorium on an asteroid as soon as possible. The expansion of Space Corporation into the terraforming phase will require a large quantity of nuclear fuel. The reason there is so much nuclear fuel now is because we have the short-term weapon stockpiles of the Cold War that we are taking apart. This may not be the case when we go to mine Ceres.

The more ships and smelters we have, the more money Space Corporation can make. As a result, construction on more ships and other equipment will begin as soon as the first mining expedition returns. The corporation will save all data regarding any issues with these machines, but planned significant changes will only be made every ten years and to as many items as possible.

Phase III: Resource-Specific Mining

This phase is identical to the second phase, but it will be different for accounting reasons. Phase II is where all of the money is made. It is where the global economy is shifted away from China, pollution, and expensive metal to the stars, free societies, and cheap metal.

Phase III exists for elements that are never brought back to Earth. In particular, we will need nickel-iron, lithium, silicon, aluminum, thorium, and uranium. We will need three million tons of lithium for our eleven thousand fusion devices. We will also need millions of tons of aluminum to build the thousands of aluminum reflectors needed to move ice and heat Mars. The uranium and thorium needed for the first few operations can be flown up from Earth, but once we have thousands of vehicles operating on Ceres, we will need hundreds of tons of radioactive fuel. In order to reduce cost, and the risk of a crashing on Earth with nuclear fuel on board, we will plan on getting all of our nuclear fuel from space by year ten.

Our initial reconnaissance will identify areas with marketable amounts of uranium, and by year ten, we will have the money to go anywhere and mine it. We should be able to find all the radioactive material we need in the main belt. Of course, whenever possible, we will integrate this effort with profitable operations. Eventually we may have a uranium mine on 16 Psyche, an iridium mine on 21 Lutetia, a neodymium mine on Vesta, etc.

WASTE-EFFICIENT DESIGNS

The interesting characteristic of nuclear fuel is that it generates heat all the time, not only when humans press the on button. On Earth, most natural energy from atomic fuel is wasted. Nuclear reactors artificially increase the output of nuclear fuel, and this is harnessed for energy. Space Corporation will make frequent use of radio thermoelectric generators. These are used on a small scale on Earth, but they will be essential for our expansion into space. Basically when heat is applied to metal electrons are emitted. This creates a low-intensity long duration supply of electricity.

Radioactive fuel is always emitting energy; it can be slowed or sped up, but not stopped. We can design all of our equipment to slowly make hydrogen propellant during the long periods of time the equipment is floating between asteroids. Even when nuclear fuel cells run so low it's not able to power a space vehicle we can plug them into a propellant station to slowly contribute energy over decades.

On Earth, we spend a huge amount of money to buffer the thermal energy out of nuclear reactors in order to make them safer. Most nuclear disasters, or near disasters, on Earth are from problems with a nuclear reactor's cooling system, since nuclear fuel is self-heating. In space, there are no living things for us to worry about, so we are more concerned with making a profit off the energy that we do bring into space.

GRAVITY CREW—HEALTH

A profitable business is one that takes care of its employees. The various national space programs have accomplished a great deal in the last seventy years, but these organizations are run along military lines, where the crew is expected to endure considerable physical hardship. Being exposed to low or zero gravity for a long time has many annoying health effects, if not even potentially serious ones. After long-duration stays on Soviet space stations, it was not uncommon for cosmonauts to be carried out of their landing craft. We are not calling on brave men and women to endure hardship out of patriotism or science, however noble an idea that is. I have never seen a hotel advertisement that extolled how uncomfortable or dangerous their establishment was and then offered to pay their customers a comfortable salary for visiting them. I want the universe to look more like Manhattan and less like the Maginot Line.

Robert Zubrin wrote in his book *The Case for Mars* that a HAB module could be rotated around another object on a sufficiently long tether to simulate Earth gravity. Of course we'll plan on making this standard for all Space Corporation missions, whether mining an asteroid or traveling to a new planet. We will not be landing any humans on Eros 433; humans will simply be on hand to pilot the unmanned space vehicles. If some of the crew wish to go down to the Erotian surface and take the greatest Facebook profile shots in the history of humankind, they will be permitted. They just have to clock out first.

The crew will be trained and equipped to function in space under zero gravity, but this will be an emergency measure, and we'll plan on avoiding it. It is important that astronauts or tourists have a pleasant experience traveling in space. This equates to better retention and higher productivity for the company workers and eventually more colonists and workers for Space Corporation activity. We'd also like to streamline our technology for other uses, and it would be wonderful if our designs were purchased by five-star (pun intended) space hotels because they were that comfortable.

Currently, astronauts need time to recuperate after returning from zero gravity. This reacclimatization period gets longer as humans spend more and more time in space. While the gravity on Mars is less than that of Earth, we would not want the first human step on another world to be an awkward stumble.

In 1973, the United States operated a space station called Skylab. One of the interesting health effects that we learned from this mission was that most circulatory and conditioning problems that the US astronauts experienced were because the body is designed to fight gravity. That's mostly what your legs do. In space, legs are basically not used at all. Not only is this bad for the leg muscles, but even the leg bones start to lose calcium. The body practically forgets your legs are there (in the same way that an athlete who stops conditioning her muscles slowly loses strength). These affects are reversible for the relatively short flights that we have been taking, but as the distance and time go up, the effects will also.

As strange an idea as it is, legs, couches, and chairs are about as useful for people in space as a parachute would be for a sentient fish. Without gravity, your legs perform no function at all. Astronauts actually complained that their legs frequently bumped into things. NASA administrators honestly recommended sending double amputees into space.[9] A double-amputee astronaut would consume significantly less food, space, and oxygen. Their circulatory system would also function much better than a normal person. Under zero gravity, blood flow to a person's legs is reduced (since the heart still thinks it must pull the blood up against gravity).

For cardiovascular fitness in space, the astronaut workout could be designed for arms, which would actually be used, rather than legs. The bicycle that was used on Skylab for the crew to exercise was unwieldy without the help of gravity to keep one's feet on the bike—the astronauts sometimes floated upside down and used their hands to spin the pedals!

While it is nothing that I am happy about, America currently has many double amputees as a result of our recent wars in Iraq and Afghanistan. Many of these soldiers have prosthetic limbs and have learned to function very well. Some single-limb amputees have even volunteered to return to combat with their prosthetic limbs. The missions to the Martian and Venusian surface would have to be reserved for astronauts in full health, but if mining expeditions were launched without any artificial gravity, I've no doubt we could find plenty of physically fit, double-amputee combat veterans willing to participate.

---

9   *Proceedings of the Skylab Life Sciences Symposium*, pp. 818, 831–832, 834–835, and 842–844.

CHAPTER 6
Atomic Ice Mining

INTRODUCTION

I know that much more attention has been paid to terraforming Mars rather than Venus. I have a radically different approach to these matters, and I found that by using fusion devices, ice blade rigs, and ballista-2s, we can terraform Venus first. Ironically, none of the crucial activity for terraforming Venus will occur physically on Venus. It will occur on Ceres and at the Venusian LaGrange 1 point. We are not atomically mining Venus or Mars; we are atomically mining *for* Venus and Mars.

Atomic Ice Mining Capability

While the small amounts of ice needed for human consumption in space will be provided by the ice asteroids already mentioned, eventually we'll have to go into the solar system and get ice on a planetary scale. In order to do this, we will do the following:

1) Mass-produce a large number of more massive and powerful ballistae in space (ballista-2s). Since these will never be launched from Earth, they can weigh hundreds or thousands of tons and employ nuclear reactors of one hundred or one thousand megawatts.

2) We will take advantage of the atomic weapons left over from the Cold War to make atomic devices and fusion devices. These will basically be high-yield atomic or hydrogen bombs capable of splitting large asteroids in half or excavating thousands of cubic kilometers of ice from the surface of Ceres. The atomic devices will yield less than five hundred kilotons, and the fusion devices will have yields roughly 6700 Megatons for Ceres, various other yields for different asteroids and comets.

3) We will build machines called ice blade rigs and TESR rigs. These will be used to cut ice out from asteroids and will make excavating them with a fusion device far more accurate and efficient. These will also be made completely in space.

These three sets of technology employed together will allow us to cut out large sections of ice, blast them free, and then guide them into their targets with ballistae-2s.

To be clear, a fusion device is similar to a nuclear bomb, but it will not be hardened against electronic countermeasures; it will be assembled by representatives of different nation-states, and most of the components will be made in space. We can also design them so that they only work when given an authorization code from Earth (which prevents the devices from being used for anything else). We can also design a system where the device can only be operated 100 million miles from Earth. Many military weapons have minimum safe arming distances, protected zones where artillery fire computations won't work, or radio beacons that prevent firing on friendly forces. If we can do this for a small military base or a blackhawk helicopter putting a 100 million mile no-fire zone around our homeworld for fusion devices would actually be quite simple. I do not use the word *bomb* to describe these devices, because they are not designed with a military purpose. This is a civil engineering project, and civil engineers are not bombers. Anyone who has had a machine *designed to hurt people* operated at, near, or by them will appreciate this distinction.

While there are several moons and asteroids in the solar system with abundant ice, I have focused on the dwarf planet Ceres due to its size, ice content, and relative proximity to the inner solar system. The Thermatid family of asteroids, the Cybele group, and Saturn's moon Iapetus are also reasonable secondary locations.

The crust of Ceres (one-quarter the mass of the entire planetoid) is ice. This single source is three quarters of all the ice needed to terraform Venus. Saturn's moon Iapetus has five times the ice needed to terraform Venus, but it's much farther away so it is a secondary source. There are thousands of objects closer than Iapetus for us to find the rest of the ice we need.

While explosions are naturally spherical, engineers can arrange explosive charges to render very specific effects on the matter they are trying to alter. In the US Army, engineers are trained on how to use explosives to make a tree fall on a specific side of a road. Some explosions are designed to destroy a building, other kinds destroy tanks. Similarly, building demolitions are frequently done by civil engineers in the middle of large cities with no damage to any nearby structures. By precisely cutting the ice out of the planetoid before we use a fusion device, it will be more similar to an ice projectile being fired with nuclear gunpowder than an explosion throwing debris in every direction.

Ceres can be reached from Earth in six months by NTRO. (It orbits the sun between Mars and Jupiter.) With one of the more advanced forms of propulsion (VASIMIR or NSWTR), we could lower this time and save a fortune on transporting the crew.

The NTRO on the Katas and the ballista will generate profit in the distances close to Earth. We can also use them to fly unmanned equipment out to distant places. If we have to, we can fly to Ceres with NTROs, but we'll have to get VASIMIR or NSWTR working before year ten to efficiently send humans to Chiron, Okyrhoe, or Echeklus. Even for Mars and Ceres this would turn a six month or three month NTRO mission into a 4.5 month or 6 week VASIMIR mission. The windfall of profits that start to come in after year eight will ensure the financial resources to fund research, construction, maintenance and fuel for more advanced propulsion systems.

Any new propulsion systems will be nested with all of our existing technology. One ESR can power one NTRO. One ESR can power one NTRO. VASIMIR and NSWTR will probably need more or different power sources. Either they can use multiple ESR-1s, or we can develop an ESR-2 that can work with older propulsion systems on a lower power setting. Liquid core and gas core NTROs are also a viable option. There will be as many interoperable pieces as possible. The VASIMIR and NSWTR engines will also be designed with the same connectors as all other equipment.

Ceres has a mass of 9 E+20 kilograms. One quarter of this mass is water ice on the surface, or roughly 2.2 E+20 kilograms. We need 3.2 E+ 20 kg of ice for Venus, so this is a good start.

ATOMIC ICE MINING PREPARATION

The position of Ceres fluctuates from 2.5 to 3 astronomical units from the sun (each astronomical unit is 150 million kilometers). Earth is pretty steady at 1 astronomical unit, so at closest approach Ceres is roughly 1.5 AU from Earth. The position of Mars is 1.4 to 1.7 astronomical units. For the ease of calculation, I am using the average measurement of 2.7 astronomical units for Ceres and 1.5 astronomical units for Mars, requiring the megapili to traverse a distance of 1.2 astronomical units relative to the solar plain. Venus is located 0.7 astronomical units from the sun, so the last leg of a megapili would be a distance of 0.8 astronomical units (between Mars and Venus). My estimate is based on 1 km/s sunward after leaving Ceres, then accelerating by another 1 km/s sunward after the gravity assist in Martian orbit. Both of these velocities can be increased if it suits us by increasing fusion device yield or the angle/distance of the gravity assist.

It will take seven years for the objects to reach Mars, and another three after that to reach Venus. The Venusian leg could probably be made faster than this, but we'll assume two kilometers per second. Past Mars, the effect of sunlight on ice will become a big problem. We will carefully set all of the courses and speeds for the megapili so that they will be flying in a line at a constant rate of speed. All of these flight paths will be planned in such a way that the megapili only approach Earth's orbit while Earth is on the opposite side of the sun.

As we cut each Megapili from Ceres we will ensure that the portions being cut are ice, not rock. We will accomplish this by sending a Tunneling ESR down the entire depth of the proposed object, equipped with a GRS. This will let us know how much rock is present along the proposed Ice-blade seem. It will not be an issue if a small amount of rock is present in the megapili when it is excavated from the Cerean surface. There will only be an issue if the rock is present along the blast seem we are melting to excavate the object. It is much simpler to melt ice and then direct the steam out of the way than it is to melt rock or iron and direct the vaporized rock or iron out of the way. The ice blades will be capable of digging through a small amount of rock, but we will prefer to slightly alter the shape or position of each megapili to avoid cutting through such places. It is likely that the entire Cerean mantle for a depth of one hundred kilometers is pure ice, with only a thin covering of rocks on the very surface. We can use the ice blade rigs and the steel support beams to dig a small trench along this thin covering of rock (depending on the depth). If the TESR exploratory tunnel confirms that there is minimal rock along the intended excavation route, we will put a radio tag at the bottom and use this to emplace all of the ice blade rigs around it.

The ice blade rigs will take much more time than using the fusion devices for excavation. We will start the ice blade rigs gradually over several years. In this way, three-fifths of all the work will be done before the first device detonates. When the time comes to excavate the megapili, a TESR rig will be used to insert a fusion secondary device at the bottom of the exploratory tunnel. All other equipment will be cleared out of the way (in case of an accident), and the fusion igniter will be added, making the device fully operational. We'll wait until the rotation of Ceres lines up with Mars and when the Ice-Blade rigs have all attached aluminum reflectors to the sunward side of each megapili.

The force of the blast will push the megapili off of Ceres, and the ballista-2s will help it escape from Cerean gravity at the path and speed we want. All of the first detonations will occur along the Cerean equator, where escape velocity is only 0.41 kilometers per second instead of the 0.51 kilometers per second elsewhere on the planetoid. We will conduct excavations from the equator toward the poles and then back at the equator once the first layer of crust has been removed.

The ballista-2s will spend a short time making minor course adjustments to the megapili going to Mars. I'm confident we can become good enough at cutting ice and releasing it with fusion devices that no course corrections with ballista-2s will be needed. We will quickly perfect this process to such a degree that the ballista-2s will only be on hand as an emergency. Bullets are fired out of guns and strike their targets without the aid of hundreds of flying robots adjusting their path.

While cutting the ice out before the blast will help to keep the megapili in controllable pieces, there will be a few stray pieces of ice that come out in the same general direction. This will be part of the pili train (or megapili train in this case). It is easy to alter the orbit of one massive object, but it is impossible or very difficult to do this for ten thousand smaller objects simultaneously. Since the initial target for the object is Mars, any small pieces that are not guided into a gravity assist will be caught in Mars orbit or fall to the Martian surface. We may even want to direct some of these objects into stable Martian orbits so that we can use them later. Since Ceres is such a huge distance from Earth and blast seems will prevent all but the smallest debris problems, this poses no impact risk to Earth. We will lose some mass from this process, but we should be able to divert lost ice to Mars (which also needs ice).

The gravity of the megapili will pull many of these objects along. If something is big enough to be a concern, any large pieces of debris floating near the megapili (dozens or hundreds of meters in diameter) will be intentionally pushed into a Mars intercept. Even the slightest nudge from a space vehicle will be able to do this (all pieces are already going in the general direction of Mars). We are going to have to move these megapili very close to Mars to conduct a gravity assist. Our ballista-2s will be attached to each magapili to make minor course adjustments. They will not be attached to nearby floating debris. While each megapili will have only a small amount of debris floating nearby 1% of mass from 5500 megapili will be enough to make an ocean on Mars.

Ballista-2s and smaller space vehicles will spend several hours adjusting the trajectory of Megapili and surrounding debris. These objects will coast for years, so the energy cost of adjusting the flight paths is lowest moments after the megapili is excavated. This will ensure we don't have loose debris in the inner solar system, as well as adding water that Mars badly needs. At this point in the terraforming process the Martian atmosphere is very thin. It lacks the density and heat needed to split water. The ice will simply melt into water vapor, and then eventually condense on the ground and freeze.

There will be careful economic considerations for Space Corporation to determine how many ice blade machines they want versus how many human missions they are willing to send out to Ceres. I have provided all of these numbers as a basis to convey an idea. Any of these numbers can be adjusted (more machines and fusion devices, for fewer ballistae, with more time, etc.).

FIRST ATOMIC MINING EXPEDITION

All atomic mining operations will be nested throughout the entire plan of Space Corporation. The first few items made from derelict space vehicles will be things useful for infrastructure (for Green Station 1, Propellant Station 1, and HAB shields).

As mining of asteroids swings into full production, the quantity of unmarketable metal will drastically go up. During the first two years, the excess metal from asteroids will go into making technology with short-term profits (ballista-1s and 2s, asteroid torches, smelter components, more propellant and green stations, etc.). By year ten, we will have a large enough fleet of space vehicles to produce the entire Earth's demand of rare earth metals and gold (approximately 120,000 tons per year). Practically speaking, as the price goes down, demand will increase, so it could be much more. This will yield roughly twelve million tons of metal by-products per year. At a minimum, the objects made from the metal by-products from this stage will be the following:

- 200 ice blade rigs
- 400 conveyor stack vehicles
- 610 track assemblies
- 610 corvus attachments
- 10 TESR rigs
- 100 ballista-2s
- 7 HAB shields
- 2 megagreen stations
- 2 megaproppellant stations
- 100,000 support beams, each one centimeter by ten centimeters by one hundred meters in diameter
- 3,500 Aluminum Reflector 1s (each 60 km x 60 km, as thin as possible)
- 11,000 fusion secondary devices (made of lithium and iron)

In space it is easier to move a low-mass manufacturing center then it is to move a larger mass of finished manufactured goods. We will try to manufacture these items as close to the surface of Ceres as possible. It would not be difficult to bring down a small metallic asteroid on Ceres (the gravity is low enough that it would be comparable to landing an airplane on the surface without landing gear). We can then land several smelters near this site and anchor them to the surface with corvi. We can use smelters to manufacture all of this equipment on the surface of Ceres so there is no transportation cost. Once we start atomically mining Ceres we can move the smelters off the surface. We will also need a large amount of nuclear fuel for this part of the plan so we will have to have a uranium or thorium mine operating at the same time.

Some low mass objects (like computer processors, sensors) may still be made on Earth, but the more things we can make in space, the better. The preliminary work of ice mining can be done with robots and small human crews, before any of the fusion devices arrive. As the machines are produced, we will start a few practice megapili removals on the opposite side of Ceres from our manufacturing center. Starting in year twelve, we will begin diagnostic work on all of our equipment to ensure our designs, training and software really work, or to make minor improvements.

By year fourteen, the ice blade rigs, TESR rigs, support beam stacks, and one mega-green station will be in Cerean orbit. We will be making the ballista-2s on a nearby (low gravity) asteroid and flying them to Ceres.

The TESRs will dig/melt down to what I call the Litho-Glacis (LG) line. This is where almost pure ice on the crust of an asteroid meets the rocky mantle. In reality, this is more of a zone rather than a line. The ratio will probably change from 99 percent ice and 1 percent rock to 90 percent ice and 10 percent rock, and eventually all the way down to solid rock. If possible, we only want to export pure ice to Venus. Based on the other sources of ice available, we may go down to larger proportions (10 percent rock might be okay). The design of ice-core machines will allow them to melt through solid rock, and we will just have to increase the temperature on the ice blade to its maximum level. Of course it's easier to melt through ice, and we will always plan to avoid metal or rock when possible.

Each expedition will use the same vehicles, but the crews will be changed out between expeditions. The fusion devices represent a total figure. We will only send a proportionate number of devices out with each crew rotation.

When the first equipment arrives in year twelve, we will start work on a few of the megapili. We will be very slow and methodical with our first excavations so we can make changes to our equipment if needed. This will allow us to verify that our designs are good, as well as correct any bugs that we notice during our trial run. We can have our production capability going to the simple items until our prototypes have been verified. Once the exploratory expedition in year twelve has concluded successfully, we will keep adding more equipment and personnel, while rotating the crews.

Most of the time consumed by atomic mining will be cutting the ice free from Ceres. Aside from a few devices used for our test operations in year twelve, none of the fusion devices or military personnel will be needed until year fifteen. When the fusion igniters are launched from Earth, the nation that contributes them will send a team of military service members with their devices. They can even come in a separate Hab from the civilian mining personnel. Nuclear weapons on Earth are maintained by the armed forces of any government that owns atomic weapons, so there will be no issue of compromising secrets or taking a security risk with citizens of a foreign power.

Since these are not weapons but mining devices, we could probably have all of the states participating agree to one basic design. In this way, there would be no risk of Americans looking at Russian technology, or vice versa. All of the technology would be shared (and separate from purely military technology). The tornado fighter was jointly built by Italy, Germany, and the United Kingdom, for example. Britain also helped develop the trident missile with the United States (the UGM-113 mentioned earlier in the book).

It would be nice if the United States, Russia, and many other nations exchanged atomic weapons technology to Space Corporation for Stock shares. I've built this plan in such a way that America could do it alone. The United States has thirteen thousand plutonium bomb cores that could be easily turned into fusion devices. I use the number 11,000 in this book on the assumption that some of those cores will go bad by the time we are ready to put them into Fusion Igniters. Not to beat a dead horse, but not one of these plutonium cores is attached to an actual functioning weapon, so this would have no impact on American national defense.

CEREAN MEGAPILI

The term *megapili* is used to denote any man-made or man-directed object that is larger than one kilometer in diameter. In this section, I will address my primary plan for getting ice to Venus and removing megapili of ice from the crust of Ceres. We can generate megapili of various different shapes and sizes from all over the solar system, but the largest amount will come from Ceres. The Cerean megapili will be shaped like an upside-down pyramid (with the base on the surface and the sides going down). It will have a base of fifty seven kilometers and a height (or in this case, depth) of fifty seven kilometers.

The pyramid will be excavated using ice blade machines. The gravity of Ceres will keep the megapili on the Cerean surface even after the sides of the pyramid are completely melted out of the crust. Since we are melting through ice and around rock there will be slight variations in size and shape, 57 kilometers is merely an average.

Below is the equation for mass and volume:

3.2E+20 kg /5500 megapili = 5.8 E+16 kg per megapili

$1m^3$(ice) = 910 kg

1 km^3 of ice = 5.8 E+16 kg/ 9.1 E+11kg = 63,736 km^3

Volume$_{(megapili)}$ = 63,736 km^3

Volume$_{(pyramid)}$=1/3*b1*b2*h

Base (and height) = 57 km

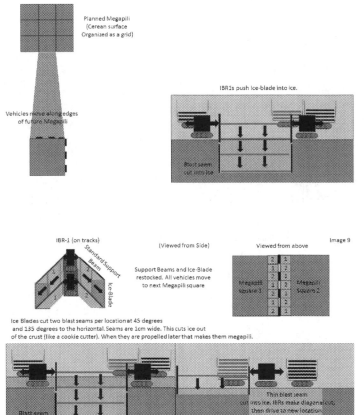

Image 8

Planned Megapili
(Cerean surface
Organized as a grid)

Vehicles move along edges
of future Megapili

IBR1s push ice-blade into ice.

Blast seem
cut into ice

IBR-1 (on tracks)

Standard Support Beam

Ice-Blade

(Viewed from Side)

Support Beams and Ice-Blade
restocked. All vehicles move
to next Megapili square

Viewed from above

Image 9

Megapili
square 1

Megapili
Square 2

Ice Blades cut two blast seams per location at 45 degrees
and 135 degrees to the horizontal. Seams are 1cm wide. This cuts ice out
of the crust (like a cookie cutter). When they are propelled later that makes them megapili.

Blast seem
cut into ice

Then blast seam
cut into ice. IBRs make diagonal cut,
then drive to new location

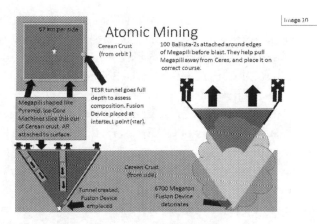

Atomic Mining

Cerean Crust (from orbit)

100 Ballista-2s attached around edges of Megapili before blast. They help pull Megapili away from Ceres, and place it on correct course.

57 km per side

TESR tunnel goes full depth to assess composition. Fusion Device placed at intersect point (star).

Megapili shaped like Pyramid. Ice-Core Machines slice this out of Cerean crust. AR attached to surface.

Cerean Crust (from side)

Tunnel created, Fusion Device emplaced

6700 Megaton Fusion Device detonates

Image 10

Dr. Birch estimated that 4E+19 kilograms of hydrogen would be needed to convert the CO_2 in the Venusian atmosphere into solid carbon and water via the Bosch reaction.

At 2,200°C, 3 percent of water disassociates into hydrogen and oxygen. At 3,000°C, 50 percent of water disassociates. At 6,000°C, 100 percent of water disassociates. These temperatures are frequently reached by objects entering Earth's atmosphere. The Venusian atmosphere is ninety-six times denser than Earth's. It is also at an average surface temperature of 450°C, where the average on Earth is only 60°C. Venera 4, the first man-made object to survive entry into the Venusian atmosphere, was built to withstand an atmospheric entry temperature of 11,000°C (1.83 times what is needed). There were several previous probes that burned up in the Venusian atmosphere (made to lower heat tolerances). Entry temperatures are affected by the angle of the object as well as the velocity of entry, but 10,000°C seems like a reasonable estimate. We will be temporarily making the Martian atmosphere hot and thick enough to bring about a 6,000°C entry temperature. The Bosch reaction itself produces heat, so once we reach this level all of the subsequent reactions will be even easier to pull off. We will break each megapili into pre-cut pieces to ensure they melt before reaching the surface.

After the hydrogen and oxygen cool from a warm 10,000°C to a chilly 730°C, they will start the Bosch reaction. The Bosch reaction will eliminate all but a trace amount of CO_2 from the Venusian atmosphere, replacing it with water and oxygen. To be fair, as the quantity of CO_2 becomes lower, it will be harder to bring this reaction about. Nonetheless, even if we had to increase the amount of ice imported (as the efficiency of the reaction decreases), we can still make enormous progress in cooling the planet. We can certainly remove a large enough portion of CO_2 so that plants and bacteria can grow there.

Ironically we need Venus to be hot during terraforming. The Bosch reaction requires a huge amount of heat on a huge amount of ice. We will wait until all of the ice has been imported into Venus before securing the sun-blocking parts of the VL1 object. We will build the support structure for the Helmholtz coil and manufacture the thin solar panels in another location. After all of the ice we need is added, we will attach the sunlight blocking parts to the VL1 object and safely cool our new planet.

ICE BLADE OPERATIONS

As mentioned in chapter 2, the primary method by which all of the ice will be cut out of Ceres is with ice blade machines. The metal at the bottom of the blade will be heated by an electric current coming down from a nuclear reactor on the surface. There will be tiny exhaust tubes made from insulated material and artificially heated so that evaporated steam can move up the blade and be placed on the surface as solid ice (CPP-2).

Every ice blade rig will need support beams to extend the actual ice blade deep into the ice. These support beams will be used for many other things throughout this book, so I just refer to them as support beams. These will be made out of the most common metal available to us (probably nickel-iron). They will be ten centimeters wide, one hundred meters long, and one centimeter thick. The ice blade will have one support beam on each side.

We will also have simple standardized tethers, which are made out of the strongest metal alloy we can mass-produce in space. If we can use lighter and stronger material than metal, we will do that as well. These tethers will be used by the ballista-2s to bring in any equipment not made on Ceres. The same tethers will also be used to tow the megapili behind the ballista-2s after the megapili are blasted free.

Finally, the standard support beams are transported on conveyor support stacks. This is a conveyor belt mounted on a track that feeds support beams into the ice blade rig. Each piece of equipment can move independently, but when they arrive at their location, they can interconnect to ensure sufficient support beams for ice blade operations. These support stacks can be parked next to each other to form conveyor belts of support beams for whatever size is needed. The ice blade rig is designed to remove the support beams from the stacks, place them vertically, and then push them down on top of the existing beams. When the beams are pulled back up, the same process is employed but backward.

All of the "1" designations on this equipment will be configured for work on the surface of Ceres. These are the same items, but they are fitted with a module called a track assembly. This gives them tracks, like a tractor or tank, and allows them to drive over rough terrain on the Cerean surface. When we want to move these objects off of Ceres, we can simply detach the tracks, tether the ice blade rigs to nearby ballista, and then lift them off the dwarf planet. Vehicle suspension, track design, and power consumption will be based off of low Cerean gravity (about 5% Earth norm).

The "2" series of equipment will have a corvus for securely anchoring any Space Corporation equipment into an ice object. This is a metal anchor that can be heated by an SPU and then melted into a body of ice. The corvus will connect to the same spot that a track assembly does, so all the equipment will only need one attachment to transition from Ceres to a zero-gravity asteroid. There may be a few other pieces added as well. The support stacks can interconnect with the Ice-Blade Rigs to stay anchored when they are in a low gravity environment. We can also attach corvi to the support stacks when needed.

While Ceres has no atmosphere, and only one-twentieth the gravity of Earth, it does still have gravity. Gravity means friction, which means loss of energy, which means loss of money. All manufacturing and transportation companies on Earth pay a lot of money due to friction caused by gravity. It's literally holding all poor people down. All possible activity for Space Corporation will be done in zero gravity to keep costs low.

The TESR rigs will be built on the surface of Ceres. When the future boundary for a megapili is planned, we will call that part of the surface a megapili square. Each megapili square will start with a TESR rig melting an exploratory tunnel from the surface down to the bottom of the megapili. While we are melting this tunnel we will be using a gravimeter and ground penetrating radar to scan the surrounding ice. The deeper we go we can even send different wavelengths of electromagnetic energy through the ice to sensors on the surface (the same way an X-ray works). This can let us know what ice or rock is above and around the exploratory tunnel.

If we find the ice and rock all the way down to be satisfactory we will plan all of the ice blade rigs around the exploratory tunnel. If we don't like the results we may dig another tunnel to identify a deposit of rock or metal that we need to avoid. After the Ice blade rigs are finished the same exploratory tunnel will be used to move the fusion device into position.

Once the location has been established, we can place small, low-frequency radio transmitters at the bottom of the hole. Since we are basically cutting out an upside-down pyramid, the ice blade rigs will drive into the shape of an L. Eventually all the vehicles will move to the other side of this L, cutting out all 4 sides of a square. The exploratory tunnel will start on the surface at the corner of the square, then go down 57 km until it reaches the center point of the square (if viewed through the ice on the surface).

Each Ice Blade rig will have support beam stack vehicles next to it so that it can extend the ice blade as deep into the ice as we need. All of the stack vehicles will have an SPU that allows them to drive around the low gravity surface. The IBR-1s will have an ESR both for moving and for generating the heat needed to melt through the ice. All vehicles will use the same set of tracks, though the ice blade rig will need multiple tracks, the support beam stack vehicles only one.

We will also bring one hundred corvi down to the sides of the pyramid and emplace those with ballista-2s. Each corvus will have an SPU for its own power. We will attach tethers to each corvus, and that is how the ballista-2s will attach to the megapili when any changes of direction are being made.

To prevent the ice blades from running afoul of each other while they are making their cuts, each team of one hundred ice blade rigs will cut in two different directions. The blades themselves will be 600 meters wide, but the machines will only be placed at 560-meter intervals.

Fifty machines will cut from the surface to the intersect point for the first grid square, and fifty will cut from the surface to the grid 90 degrees from the first one. When all the machines are done making this cut, they will pull the ice blades out of the ice, rotate them 90 degrees, and make a cut in the opposite direction. In this way we can keep the blades lined up without them running into each other.

Each team of ice blade rigs will make two different cuts and then drive to a new location, lined up the same way they were. Each machine will have a specific length of support beams, and their own place in line. Machines in the center must melt down the full 80 km, where machines at the sides will have only a few meters to melt. The average depth for all machines will be 40 km.

We will then manufacture an aluminum reflector and cover the surface of each megapili. We can melt the edges of the aluminum reflector into the ice to hold each one in place. Once all preparations are complete, and all of the ice blades are out of the way a TESR rig will be used to place a Fusion device at the bottom of the exploratory tunnel. The bottom of this tunnel is the point at which all ice blades intersect. A release of energy at this point will have nowhere to go except through the blast seems, propelling the object into space. Of course we will test this process on smaller asteroids or even a few blasts on Ceres before we start full scale excavation.

By using this method, the entire surface of Ceres will be excavated. The rough math for this will be 852 objects on the original surface. The second layer is also 852 (the second layer resembles serrations compared with the first layer, they are interlocking pyramids). The third and fourth layers will be about 600 objects, each. Finally the 5th and 6th layer will be about 300 objects each. There is a huge amount of ice on the surface of Ceres, we will keep digging down and excavating the ice until it is all gone. In total Ceres will give us between 58% and 68% of the ice we need to Terraform Venus. This will be 3,500 out of 5,500 megapili we need for Venus. There are thousands of moons, comets, asteroids and Kuiper belt objects for us to find the other 2,000 megapili.

USE OF FUSION DEVICES

The fusion devices themselves will consist of two components:

1) The fusion igniter, basically an atomic device (four hundred to five hundred kilotons in yield). This will weigh roughly one or two tons and be closely guarded by the contributing state's military. These will be manufactured on earth and flown up in separate pieces, to be assembled near their objective.

2) A bulky piece called a fusion secondary device. This will be made up of concentric layers of metal and lithium. Once the first stage is compressed by the igniter, it creates a powerful fusion reaction. This reaction in turn compresses the next stage, etc. Three stage devices were designed that would have yields in the thousands of megatons fifty years ago. There is no known size limit for this design.

Assuming 5,500 megapili at 1.45E+16 kilograms each, we will need fusion devices with a yield of 6,750 megatons. At 9.5 megatons per ton of fusion device, this would be 710 tons (most of which is lithium and steel).

The fusion igniter is itself an atomic device (about 500 kilotons). It can be disassembled and flown up in pieces, so there is no risk of an accidental detonation. The secondary fusion device makes up virtually all of the mass (708 out of 710 tons). Not only is it harmless by itself, but it's not even expensive. On Earth, these commodities are worth about nine dollars per kilogram.

The plutonium core and the complex detonation equipment will be flown up from Earth and designed in partnership between that particular government and Space Corporation. If the governments involved want complete control of the fusion device project that would be fine. It will be in our contract that they only get paid if their device works.

Hopefully each government's desire for profit (and fear of embarrassment) will ensure that these machines work. The two devices will work together, but they don't need to be made together. A good analogy is placing a key into a lock. The lock is much more massive, but the key fits inside.

The military personnel will fly with these systems all the way to the destination, and the detonation codes will be held by that government on Earth until the last second. There's no rational risk of these weapons being used to ill purpose; after all, we have thousands of these things pointed at each other *on earth right now.* The crew will also be on hand to conduct inspections, perform maintenance, and eventually activate the devices.

During asteroid mining, we will identify large amounts of nickel and lithium. Lithium has some use for batteries and electronics, but it is not expensive enough to export to Earth (nine dollars per kilogram). We will keep a small amount of lithium for the electronics we make in space, but we will also make blocks of lithium for use in our fusion devices.

The wonderful thing about nuclear fusion is that the mass-to-power ratio defies the wildest dreams of science-fiction authors. The Tsar Bomba made by the Soviet Union during the Cold War was designed to release energy of one hundred million tons of TNT, and it only weighed twenty-seven tons. That was in 1961, with communist technology. We will not need much lithium to change the solar system. When the actual igniter arrives in space, the fusion secondary device, which has most of the mass but little of the cost, will be waiting at the top of the gravity well.

At the height of the Cold War, the United States and the Soviet Union made three-stage fusion bombs. These weapons were capable of almost unlimited yields of energy. One thousand megaton yields were designed. The reason mankind never made one larger than fifty-five megatons was because we ran out of planet to blow up. Even the fifty-five megaton Tsar Bomba was designed at one hundred megatons; it was intentionally weakened in its third stage over fear of it being too powerful. This would be like designing a five-hundred-pound bomb casing but filling 250 pounds of it with tennis balls because you're testing the bomb a few blocks from your daughter's elementary school. Even in its reduced state, the Tsar Bomba damaged structures one thousand kilometers away and had a mushroom cloud so huge that the top of it left Earth's atmosphere.

This weapon's designer, Andrei Sakharov, was shocked at his own creation and led a campaign against nuclear weapons after the Tsar Bomba was detonated. Though the Soviet Union was without political liberty at this time, it seems that Sakharov's prestige secured his safety, and no bombs of this size were detonated again. A lot of people were shaken by the incredible force of these devices and their terrifying effects. They were simply too large to be used on Earth. The largest US bomb to be set off above ground was 15 megatons. My readers may recall a Chinese proverb about using a hatchet to remove a fly from a friend's head. There is no rational military application for bombs of this size on Earth.

The beautiful part of my idea is that we can harness the incredible energy of these devices, and do so one hundred million kilometers from Earth. It requires a huge amount of energy to move 3.6 E+20 kilograms of ice to Venus. By using fusion devices, virtually all of that energy comes from lithium, a cheap common element.

Based on the most advanced US fusion bomb design, a potential ratio of a 9.5-megaton yield to one metric ton of weight was deemed possible (though only bombs of ratio eight to one ton were built). We will plan the size of the TESR, proto-ballista, and fusion devices we are using so that they can all work together. While the sides of this object will be melted out to one centimeter in width, the TESR tunnel will have to be at least one meter across. We will assemble the fusion devices so that they fit in as narrow a tunnel as possible (even if they are one hundred meters long).

We will secure the rest of the ice that Venus needs from smaller asteroids, comets or moons. We may need several different sized fusion devices, but each will follow this general scheme. The US navy has adjustable yield nuclear weapons; we can certainly create adjustable yield fusion devices.

Whenever this method is used it will follow the same basic method employed on Ceres, we will just vary the size of the megapili and fusion devices based on the size of the object we are atomically mining.

The comet 2060 Chiron will be the largest zero gravity object that we will get megapili from, so I will describe that operation. We will fly a crew out to Chiron with the same equipment we used on Ceres. We will cut out one megapili and excavate it with a fusion device. We will plan the flight path so that this megapili will be propelled plutoward. Newton's third law of motion would then push Chiron sunward, collapsing it's orbit to line up with Saturn's orbit. We will then use ballista-2s and Saturn's gravity to perform a gravity assist, where we can radically change the orbit of Chiron, sending it directly to Mars. We could also do a second gravity assist around Jupiter if we needed to. While coasting to Mars we will use Ice-Blade rigs to cut it into 8 pieces, then add millions of Mtesr tunnels to each piece. This will ensure each piece breaks up properly when it arrives near Martian orbit. This same method can be used for all large asteroids, comets, and Kuiper belt objects.

ALUMINUM REFLECTORS

Aluminum is made from the naturally occurring ore bauxite. It is not hard to find on the earth's surface, I doubt it will be hard to find in the solar system. While we are bringing metal asteroids to Ceres for our manufacturing center we can also bring in several bauxite asteroids.

We can make very large rolls of aluminum on the surface. As the Ice blade rigs go around cutting out megapili we will send another small team of modified Ice blades to unroll huge sheets of aluminum foil over the entire surface. Each roll will be melted in place at the top and bottom, with most of the length of the roll left unsecure. There is no weather or gravity in space. We can use pins through the aluminum and into the ice to secure the ARs in place. We can remove these pins when each megapili is approaching its destination. We will standardize the size of the aluminum rolls with the ice blades, 600 meters across. In this way roughly 100 ice blades will be able to roll the aluminum across the top of each megapili (we can even have the same Ice blades do this as they cut the ice out and drive across the megapili to the next one). We will leave a small portion of aluminum folded over so that 100 corvi can still attach to the surface. Once the megapili's trajectory has been precisely laid in for Mars proto ballista can detach the corvi and pull the aluminum foil back over the edges. We will plan for the Aluminum Reflector to extend at least 1.5 km past the edge of the megapili. This will block light even if the megapili is not approaching the sun head on.

For the first few hours after a megapili has been excavated from the surface of Ceres, one TESR-R2 and ten MTESR-R2s will attach and cut blast tunnels and microtunnels from the outside of the megapili to the center of mass. These tunnels will ensure that a fusion device placed in the center of mass will break the object up into small pieces. We will not use this second fusion device until the megapili is just outside of the Venusian atmosphere, seven years later. Cutting blast tunnels into a megapili can alter its orbit. We'll melt the blast tunnels while it's floating toward Mars then ensure the trajectory is correct when we're done. The Aluminum foil will be laid down in strips, but all of the strips will be attached to each other at one end. When we want to remove the Aluminum Reflectors we just have to remove a few pins and push it into a gravity assist. We can attach support beams across the top to give it a rigid shape. We will fly this aluminum reflector to one of the Martian LaGrange objects, where it will be permanently put in place to add sunlight to Mars.

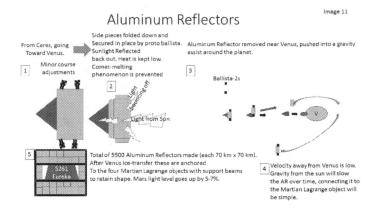

Aluminum Reflectors

Image 11

From Ceres, going Toward Venus.

Side pieces folded down and Secured in place by proto bailista. Sunlight Reflected back out. Heat is kept low. Comet-melting phenomenon is prevented

Aluminum Reflector removed near Venus, pushed into a gravity assist around the planet.

Minor course adjustments

1

2

Light bouncing off

Light from Sun

3

Ballista-2s

V

5

5261 Eureka

Total of 5500 Aluminum Reflectors made (each 70 km x 70 km). After Venus ice-transfer these are anchored To the four Martian Lagrange objects with support beams to retain shape. Mars light level goes up by 5-7%.

4 Velocity away from Venus is low. Gravity from the sun will slow the AR over time, connecting it to the Martian Lagrange object will be simple.

The Aluminum Reflectors are important because without them melted ice would disrupt their flight path, reduce their usable mass of water, and litter the inner solar system with small, annoying pieces of debris. We will manufacture 3,500 such aluminum pieces on Ceres. Wherever we decide to get the other 2,000 megapili for Venus we can manufacture aluminum at or near those locations (we can call these AR-2s).

As we pull each reflector off the megapili, we will save energy by moving each one into a gravity assist around Venus. This is actually a slight change in trajectory but it will give us plenty of energy to circle Venus and head directly for the Martian LaGrange point.

SCHEDULE FOR ATOMIC MINING OF CERES

A megapili with a base that is fifty-seven kilometers by fifty-seven kilometers will need two hundred ice blade machines, each machine using an ice blade six hundred meters wide. An ice blade is the same idea as a core sample, except that it is intentionally wide and thin, resembling the blade of a knife. By heating a small metal strip with suction for water exhaust, we will make a small cut through the ice, separating a large pyramid shape from the crust. This pyramid will still be held by the gravity of the asteroid but not by the chemical bond of the ice. As the machines push deeper and deeper into Ceres, we will need less of a melting surface (the pyramid contracts as you approach the tip). This can be accomplished by allocating a different number of support beams to the two hundred ice blade machines, with the most support beams in the center, the least on the edges, and so on.

We will manufacture two megagreen stations and two megapropellant stations near Ceres as well. There is abundant metal and ice in the area of Ceres; we only need to fly the seeds from Earth, and nuclear fuel from a different asteroid. By the time we are ready to start atomic mining we will have plenty of food for our crew and propellant for the ballista-2s to direct the megapili to Mars. One megagreen station can feed hundreds of people; we are making two just as a redundancy. The megagreen stations will be useful for morale as well, they will look like a public park designed by Salvador Dali.

The two megagreen stations in orbit around Ceres will be the type used for terraforming (each will be big enough to fit one thousand trees). We will need to cover the surface of Venus and Mars with leguminous plants, and cultivating these crops for food will be good practice when they must be grown in large numbers. The edible parts of the plants are often seeds, so whatever the astronauts don't eat can be stored for later.

All of the machines in this plan will be automated. The human crew will be there only to make specific decisions based on unforeseen events. During the considerable amount of time that Predator drones are able to fly, only a few minutes of direct human control is required. Critical functions like landing, taking off, or firing a missile are closely observed by a human being, but for the most part, these machines control themselves. This is the same concept for the ice-core machines, TESRs, ballistae, and proto-ballistae.

The crews for atomic mining will be broken down into three separate functions:

1) Ice blade operations
2) Excavation/Trajectory adjustment
3) Blast Tunnel Preparation

While functions two and three must be done together, function one can be done slowly over several years. Since all of the equipment will be remotely controlled from the HAB, we will attach all of the HABs to the green stations. This would make the astronauts feel less cramped and make it easier to transfer oxygen and food.

Ceres is so far from Earth that any radio signal will take between twelve minutes and ninety minutes. Think tanks of engineers and physicists will be of little use to us, though of course we will have experts on standby if it's something that several hours of deliberation may resolve.

Ice blade rigs and TESR rigs will be used to cut the megapili out of the Cerean surface. Fusion devices will be used to excavate the megapili, with ballista-2s on hand to make minor course adjustments, sending the megapili to Mars. The ballista-2s will use corvi to anchor to an ice object, then simply reheat the corvi when they want to leave. While the ballista-2s are adjusting the trajectory of the megapili blast tunnels will be made using TESR rigs and Mtesr rigs. Blast tunnels will perforate the megapili from the surface down to the center of mass. Years later when the object is near Venus one Fusion device in its center of mass will send energy through the Mtesr tunnels, breaking the object into 1-2 km chunks that will melt nicely in the Venusian atmosphere.

Once the megapili is excavated the ejection team will observe the megapili and line it's trajectory up with a Martian gravity assist. Since these adjustments wil be done a short distance from Ceres we can have the same group of humans adjusting several megapili with different fleets of ballista-2s. They will also move any large pieces of debris on a course to melt in the Martian atmosphere. The Ice-Blade team will wait to conduct a blast until the ejection team is back in place.

One Martian year is two Earth years, and the Cerean year is 4.6 Earth years. We can stagger the direction and speed of the megapili so that seven hundred objects can perform a gravity assist with Mars over the course of one Earth year.

When the one year of duty comes to an end, the relief crew will be on hand to work alongside the new crew for a few weeks. Each crew will travel to and from Ceres in a HAB. These vehicles will have to be traveling at very high velocity so we will make small green stations to attach to the HABs, we don't want them to dock for several months.

I'm planning a six-month journey from Earth to Ceres, one year on or near Ceres and then a six month journey back to Earth. This will not begin until the fifteenth year of Space

Corporation, so we may have VASIMIR or NSWTR working for faster movement. The fiscal benefits of faster transport will incentivize research and development. This is how jet engines, the internal combustion engine, steam power, and the wheel were invented: in expectation of profit.

GRAVITY IS HOLDING THE POOR DOWN!

It is cheaper to atomically mine an asteroid the lower it's gravity/mass is. Gravity is friction against energy and profit. If truck companies didn't have to worry about gravity they would have virtually no fuel bills, virtually no maintenance and no ramps. Space commerce will avoid gravity whenever possible, and terraforming will be no exception.

I have already highlighted how to remove ice from Ceres. Ceres is the largest single source of usable ice on our side of Jupiter, but its size can have disadvantages. The gravity of Ceres means that any mass on the surface has to reach 0.51 kilometers per second (0.41 kilometers per second from its equator) in order to escape. This can be done with fusion devices and ballista-2s, but it will be a challenge.

Let's assume we find an asteroid made out of pure ice, shaped like a one-hundred-kilometer sphere. The simplest way to get this ice to Venus will be to use Ice Blade-2s to cut the object in two sections and then place a fusion device at the seam between the two sections (still loosely connected). We will make our blast seams in a direction perpendicular to the objects orbit, directing some energy sunward and some plutoward (instead of its original lateral orbit). Once the fusion device is used, the energy will push one section of the asteroid sunward and the other section plutoward (the plutoward part becomes Newtonian ballast).

We will then have ballista-2s on hand to guide, accelerate, or slow the megapili into the precise trajectory needed to rendezvous with Venus. The plutoward megapili could also be used for terraforming, but since it's going in the wrong direction, that could take much more time. We could do a plutoward gravity assist and bring it around Jupiter or Saturn if we are patient enough.

We can also dig small pyramids in the object, just like we did on Ceres, bury a fusion device, and blast more pieces of the object to Venus. This will entail more time and more fusion devices, but it will give us a larger quantity of mass.

While the megapili is en route to Venus, we will use the TESR-2 rigs to cut blast tunnels down to the center of mass. Based on the shape, we will cut multiple blast tunnels so that our object will split and be pushed off at the velocity and trajectory we want (or close anyway). The ballista-2s will be on hand to adjust the velocity and trajectory if needed. We will also cover the sunward side of any object so that sunlight will not melt off volatile materials (Aluminum Reflector-2s).

Ceres is not big enough to provide all of the ice for Venus but we can meet this shortfall in a number of ways. The simplest would be to waste as little ice as practical on Ceres. There are also 5900 main belt comets, most of which have ice. There are millions of main belt asteroids, thousands of which have ice. We can coordinate this with terraforming Mars so that the proper composition of comets and asteroids goes to their respective planet. Even if none of these objects have the composition we need, Iapetus (moon of Saturn) is known to be almost pure ice and has a mass of 1.8 E21 kg. This is five times what Venus and Mars need. Several moons of Jupiter are known to contain a large amount of ice: Metis, Adrastea, Amalthea, and Thebe. Finally the Kuiper belt has a huge amount of ice, methane, ammonia, anything we want. We can make a specific recipe after the Katas are finished, but the Solar system has a grocery store of ingredients for our two planetary cupcakes.

FUEL SOURCES FOR ATOMIC MINING

We can get all the propellant we need from electrolysis of water in space. We already have all the plutonium that we will need for the fusion devices. The one thing missing is the large amount of uranium or thorium needed for the space vehicles involved in atomic mining. Anything launched from Earth, particularly at the beginning of Space Corporation, will be powered by uranium (more energy). As our infrastructure develops and our energy consumption goes up, we will probably switch to primarily thorium reactors in space. We may keep uranium for our ballistae and Katas (which will operate independent of refueling stations for several years at a time).

When we conduct atomic mining on Ceres, the thousands of ice blade rigs and TESRs will need enough energy that mass-producing thorium pellets in space will be a reasonable possibility. This will also allay the fears of environmentalists on Earth. No one will worry about the risk of rockets crashing full of nuclear fuel, since after year ten there won't be any Space Corporation ships near Earth with more than a few kg of nuclear fuel.

We could probably provide this uranium from the supply owned by various governments on Earth. This is certainly what we'll have to do on a small scale for the first few years before significant mining activity takes place. But the large scale of atomic mining will only be possible after nuclear fuel is refined in large quantities in space.

There is currently no real market for thorium at all. Peaceful nuclear power was developed as an accident while trying to make atomic weapons during World War II. Nuclear power plants are capable of enriching uranium (making it strong enough to use as an atomic bomb). Thorium is too weak to make effective nuclear weapons out of, and so no governments have bothered using it in power plants.

While thorium is about one-quarter of the energy of uranium, we may be able to obtain it for one-twentieth of the cost. Since all of our fuel pellets will be the same size, all nuclear fuel types will work in all machines.

One good solution is to use thorium for all of the ice blade operations and then uranium for all of the ballistae and Katas. Overall thorium has about one-quarter of the energy per mass that uranium does. One kilogram of U-235 has three million times the amount of energy as one kilogram of coal. One-quarter of that still equals 750,000 kilograms of coal.

The inner solar system has the same bulk composition as Earth. We have already discovered tiny amounts of uranium and thorium on the moon and Mars. These are two of the elements the Katas will be looking for. After the large expeditions begin launching in year eight, we will plan one of these expeditions to go to an asteroid with a good source of uranium. Uranium can be extracted by the same smelting methods I've outlined earlier in my book (though much more carefully). Eros will be a large manufacturing complex. We can comfortably import thousands of tons of uranium and thorium to Eros since its millions of miles away from Earth.

Roughly, by year eleven we'll need to have all of the equipment and raw materials at hand to mine uranium or thorium on a large scale. Once we have uranium and thorium mines up and running (likely on 16 Psyche and 21 Lutetia) we will have a huge amount of low mass fuel. We can also produce energy using solar cells, harnessing the solar wind, even using the gravity of a moon or planet. Enceladus is heated entirely from the gravitational energy of Jupiter for example. Of course, just like all of our other metals, we should be able to very cheaply process uranium or thorium from higher-grade ores than are found on Earth. A Space Corporation nuclear disaster would show up in every earth news website, but not a single earth-Geiger counter.

ATOMIC MINING CONCLUSION

The idea of using atomic weapons to send ice to Venus and Mars is so good that, even if it failed, humankind would be better off having tried. Most of the equipment and resources being retained by the US and Russian governments don't even have a military use. Unlike conventional weapons, nuclear weapons can only be used in a very limited exchange and in a very short amount of time. We have had armed conflicts in which thousands of ships and aircraft took part. We've never had a war involving three atomic weapons.

Imagine a soldier in a modern army who was ordered to carry a nineteenth-century cannon ball around with him at all times, even though no one in his army had an actual cannon to fire it. Why not sell the cannon ball for scrap? As a former member of the American military, I take my nation's security very seriously. Sadly, most citizens of atomically armed states have seen the *Godzilla* movies but have not read Herman Kahn's book *On Thermonuclear War*. There have also been antinuclear films with zombies; I remember one that had giant ants. But ask someone to explain how one kilogram of uranium has the same energy as three million kilograms of coal, and they'll feel more terror than if they were face-to-face with a giant bunny!

No soldier or well-informed citizen wants responsibility for weapons that are not even assembled. This wastes huge amounts of money and creates danger of theft or accident. Russia is in the same position as the United States, only much worse. Russian citizens have been caught selling ancillary atomic equipment to many nations, and they have far more total atomic weapons then everyone else on Earth combined.

If the US government were to use a fraction of its nonoperational atomic weapons, atomic fuel, and ballistic missiles for this purpose, we would save money *even by failing*. Hopefully this will be an honorable and profitable opportunity for Russia to cooperate on nuclear reductions, but America can safely and profitably do this alone.

I do not know what will happen to the vast array of destructive power that was created during the Cold War. It is likely that we will just awkwardly guard the warheads and spend taxpayer money to blow up the missiles. But what a unique opportunity we are allowing to rust year after year.

We can't turn chemical weapons into soap. We can't turn biologic weapons into pets. But we can, use world destroying machines to create two worlds, and I for one believe that we should.

CHAPTER 7

Terraforming Venus

Most of the difficult part of terraforming Venus is atomically mining Ceres. While we will have to remove the megapili of ice from Ceres and carefully direct the ballet of gravity that is orbital mechanics, most of the energy exists right now in little pieces of lithium. We just have to let it out at the right time and place.

I've broadly separated the steps for terraforming into three basic concepts:

1) Kinetic terraforming
2) Photonic terraforming
3) Biologic terraforming

KINETIC TERRAFORMING

Once the ice is imported into the atmosphere, it will start the Bosch reaction. The Bosch reaction is a chemical reaction between CO_2 and hydrogen that produces elemental carbon (graphite), water, and a 10 percent return on the invested heat. The hydrogen will be split off from the oxygen while the ice is entering the Venusian atmosphere between temperatures of 6,000°C and 800°C. After the hydrogen is free and starts to cool, it will go through the Bosch reaction with CO_2 already in the Venusian atmosphere. This will create solid carbon and liquid water as by-products. At first this water will form steam at the average Venusian temperature of 462°C. Even the solid carbon may be kept aloft as dust.

As more and more ice is imported, the Bosch reaction will remove CO_2 from the atmosphere, cooling the planet. This cooling will make the steam condense to clouds and eventually cause rain. The solid carbon dust will float down to the surface. We will be left with a large amount of oxygen through this reaction, but that will dissolve in the oceans and be absorbed by the (now cool) Venusian surface. Besides, I'm personally a big fan of oxygen. It's pretty much my favorite thing in the whole atmosphere.

The overall reaction is as follows:

$$CO_2(g) + 2 H_2(g) \rightarrow C(s) + 2 H_2O(g)$$

While this reaction is sped up by the presence of iron as a catalyst, the enormous heat and pressure of Venus will make a catalyst unnecessary.

I plan to import all of the ice into the atmosphere first and then setup the VL1 object to block half of the sunlight Venus currently receives. In the long term, terrestrial plants and animals will need roughly the same amount of sunlight they get on Earth, and in the short term, this will assist with cooling and terraforming the planet.

Venus also has a rotation problem; it rotates much slower than terrestrial life-forms are used to (though someone from Alaska would be able to deal with it). A Venusian day is about 117 Earth days. This was probably caused by a large asteroid impact one billion years ago, and it can be uncaused. By flying each mega pili along the axis of rotation for Venus, we can increase the spin of the planet with each object. We will break each piece up with a fusion device just before it enters the Venusian atmosphere (approximately two hundred kilometers above the surface of Venus).

We have plenty of ice in the solar system to run the Bosch reaction, so we can eventually wash all but a tiny amount of CO_2 from the Venusian atmosphere. Venus currently has about three times the nitrogen as Earth. As the atmosphere cools, this nitrogen in the atmosphere will dissolve in the oceans and in the soil.

The overall plan for kinetic terraforming is to find 5500 ice objects that are 57 km in diameter. We will use Fusion device to send them to Venus, with ballista-2s on hand to make minor course adjustments. All of the megapili will have blast tunnels going from the surface and intersecting at the center of mass. When the megapili are an hour from colliding with Venus the Aluminum Reflector will be removed and flown into a gravity assist. We will wait until each megapili has contributed angular momentum to Venus from its gravitational pull, then we will break each megapili into thousands of 1-2 km pieces.

The megapili earmarked for Venus will be excavated from year 15-23. I'm planning for a seven year movement (5 years to Mars, then a gravity assist to accelerate, and two years to

Venus). The megapili will come in at the proper trajectory to slightly increase the spin of Venus. Just before it reaches Venus the Aluminum Reflector will be removed and pushed into a gravity assist to the ML4 or ML5 point. The Megapili will have all of the blast tunnels prepared before it gets near Venus. The only thing done near Venus is minor course adjustments and using extra fusion devices if there is a malfunction.

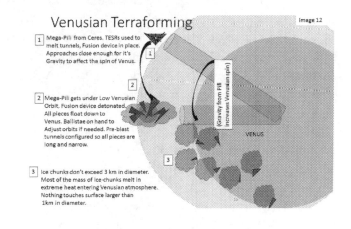

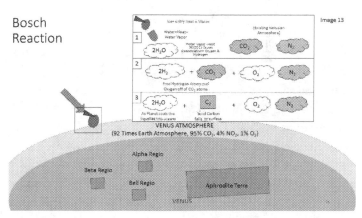

VENUSIAN PHOTONIC TERRAFORMING

Of course, Venus will not need mirrors, being close enough to the sun already, but it will need a very large Helmholtz coil (roughly eighty thousand kilometers across). There are a total of thirty-two Aten asteroids whose orbits are between Earth and Venus. One of these is 99942 Apophis (notable in the press for coming near Earth in January 2013, and once feared for a potential Earth impact in 2036). This asteroid has a mass of approximately four million tons. Assuming that 1 percent of its mass is marketable material, this leaves 3.96 million tons to be built into useful bulky components. We can use ballistae or small atomic devices to move Apophis into the VL1 point. Apophis is made out of iron and silicon, amongst various other materials. These materials are excellent for making both solar cells and an electrically conducting frame for the solar panels. We will use all of the Aten group asteroids for profitable mining. All of these asteroids are relatively close to Earth and Venus, so it will take less energy to move them, or pieces from them. In particular, the Aten group asteroids we will focus on are 2001 $CK_{32,}$ 2012 XE_{133}, and 2002 VE_{68} (all included on the Kata target list).

Relative to a thin solar shield and some hull plates there is plenty of mass amongst Aten group asteroids. These are some of the closest ones to earth, so they are likely to be a prime target for mining early on.

I realize I talk about nuclear power a great deal in this book but we will build solar infrastructure where we operate as well. I've focused on the first 40 years of my plan, when nuclear power will be needed to move moons and melt asteroids. We will set up Solar power for long term sustainability at our various space stations and surface colonies. As we make the first few panels for VL1 we can use that energy to build the rest of VL1.

We can mine metal from these objects and use excess silicon to make millions of solar panels. We'll design them to attach together in series. In the absence of gravity, they will not need any kind of support structure to remain in position.

It will be difficult to make the enormous number of panels required for photonic terraforming, but we have decades to do it. Without gravity we can move the panels around the solar system without too much difficulty. We can even use energy from the first few thousand panels to help make the rest of the panels. Even if the Aten group runs out of particular elements there are many S-type asteroids in the main belt that we can mine as well. Virtually all of the sun shield will be a very thin structure made out of solar cells.

We will have a general design worked out for this object from the beginning of our movement to space. Whenever asteroid mining is conducted on an S-type asteroid (S is for silicon), we will ensure that marketable material is removed and the unmarketable material is built into solar cells that interconnect and are of standard size. We'll need 113 million square kilometers of solar panels to block 50 percent of the sunlight reaching Venus and power the Helmholtz coil blocking 100 percent of the radiation. As long as we do a good job of planning the mining effort, it will not be difficult to make these solar panels as a waste product of removing platinum, gold, or rare earth metals.

Using an S-type asteroid for rare metals will yield 99 percent silicon as unmarketable material. As the silicon is separated out, we can add a few other elements to make it a solar panel. When mining a C-type asteroid, we will make support beams from the unmarketable mass; when mining an M-type, we can make more complex metal components.

In one year on Earth, we consume 125,000 tons of rare earth metals, gold, and platinum. If we assume that 1 percent of processed material will be marketable, that means we'll have twelve billion tons of unmarketable material for space construction each year starting in year ten.

The mining division of Space Corporation will not make solar panels with their unmarketable material because of their devotion to mankind. They will be doing it because each solar panel they make can be deducted from the 10 percent terraforming expense they have to contribute each year. They will also be doing it because once Venus is finished; it can be sold for a few trillion dollars. It will be as likely as someone hiring an accountant to get out of paying all of his or her taxes.

There are many different possible designs for this object. We will alternate the tilt of the solar cells (some toward the sun, some away) so we can neutralize the momentum imparted by solar wind to keep the object in place. We'll also have all of the solar cells feed their charge into the electromagnets that power the Helmholtz coil, or coils. The people on Venus will not be able to see it without the aid of special optics. The sun will appear larger to them, but it will be about as bright as the sun appears on Earth. The total effect is that Venus receives the same electromagnetic radiation and sunlight as earth.

We'll have a few reactors at this site to raise the energy level in the event of a solar flare. Fortunately the sun also brightens when there is a flare, so the solar cells will have slightly more energy as well. Every eleven years, the sun goes through a cycle of increased activity. We'll design the VL1 object with enough strength to survive the worst solar flares on record plus %10. After all of the ice mining operations are finished, the ice blade machines will have their ESR power plants removed, and they will provide the backup power units for the VL1 object.

We would also want the electric current fed into the Helmholtz coil to be very simple and primitive so it would not require maintenance or engineering to be maintained. The

Roman aqueducts were designed to run only by gravity— no significant maintenance was necessary. Two millennia later, some of them are still delivering water to the city of Rome near the Spanish Steps. If there were any problems with the coil, that would only result in a decrease in the quantity of current, not a catastrophic failure.

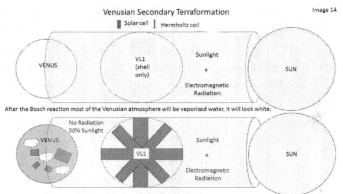

Venusian Secondary Terraformation

Image 14

Solar cell   Hermholtz coil

VENUS

VL1
(shell
only)

Sunlight
+
Electromagnetic
Radiation

SUN

After the Bosch reaction most of the Venusian atmosphere will be vaporized water, it will look white.

No Radiation
50% Sunlight

VENUS

VL1

Sunlight
+
Electromagnetic
Radiation

SUN

With sunlight and radiation reduced, this water cools and comes down as rain.

VENUSIAN BIOLOGIC TERRAFORMING

Most of the CO_2 can be removed from the Venusian atmosphere by direct importation of ice from Ceres. We will also have to convert some of the atmospheric nitrogen in the Venusian atmosphere into organic nitrogen. I do not have any amazing, high-tech way of doing this. Actually, my solution is about three billion years old: cyanobacteria.

Cyanobacteria (also called blue-green algae) played a huge role in shaping the environment of Earth. Much of the oxygen you are breathing was probably made by cyanobacteria over the course of three billion years. Atomic oxygen (O_2) is very rare in nature; Earth originally had a nitrogen/CO_2 atmosphere. Plants and cyanobacteria slowly evolved and spread across the planet, consuming the N_2 and CO_2 and then replacing it with oxygen and organic nitrogen. Eventually oxygen-breathing organisms evolved, and that's been en vogue ever since.

As a result of competition, cyanobacteria have evolved to reproduce much faster now than their ancestors. Cyanobacteria blooms can spread over areas the size of a small country in a matter of days whenever they find an unusually rich source of nitrogen to use as food. After ice is imported to Venus, the Venusian ocean will have much more nitrogen dissolved in it than Earth's ocean does. As the cyanobacteria eat the nitrogen in the ocean, more nitrogen will be sucked in from the atmosphere. Of course we will specifically pick the best cyanobacteria for this; we may even genetically engineer it to be better suited to Venus.

Cyanobacteria are spread by microscopic pores. They are so easy to spread that we will not require a green station. One or two capsules launched from Earth, if they are broken up high over the ocean to ensure a wide distribution, would cause dozens of cyanobacteria blooms. Assuming the same rate of growth as on Earth, but with near infinite food for them, they will cover every inch of the ocean in a matter of months. On Earth, algae grow even in dew on grass. It grows in canned food and in rain gutters. (Pond scum is deceased algae.)

About 80 percent of the Venusian surface will be covered in water, but we can also fixate nitrogen on land. We will have to add bacteria to the Venusian soil so that plants will take to it properly. We can coat our seeds with this bacteria before they are brought to the surface. With an infinite amount of food and space it will not take long for bacteria to spread exactly where the seeds are.

All legumes have the ability to take natural nitrogen from the soil and break it down into organic nitrogen. I used tipuana tipu trees as an example of a fast-spreading land plant, but there are hundreds of different species that are useful as well. Two in particular are peanuts and soybeans. Each peanut and each soybean is itself a seed for that plant. We can use seed canisters to distribute these over the large landmasses on Venus. We can use OPA-3s to deliver them to small islands or isolated areas. One seed canister will spread seeds out over several kilometers, but an OPA-3 can be set to a few hundred meters.

Another wonderful characteristic of these two plants is that they are capable of sexual and asexual reproduction. Sexual reproduction permits better adaptation and long-term species health. Asexual reproduction ensures species survival in temporary difficult circumstances (like colonizing Venus).

There are also leguminous grasses. Clover and alfalfa are legumes. We will probably mix multiple species in each seed canister or terraform specific islands and regions with plants based on what our customers want, or what we think is best. We will basically be building the Venusian food pyramid from the bottom up.

While Venus and Mars today have very different climates, the goal is that we make both of them earthlike. For the most part the same species should be able to spread on both planets (both will be rich in atomic nitrogen). This will allow us to maximize the use of our mega-green stations, seed canisters, and terraforming expertise.

ACTIVITY ON THE VENUSIAN SURFACE

After the asteroids have altered the Venusian atmosphere and the solar shade has reduced the temperature, we will bring a metal megapili down into an area where a freshwater river runs into the ocean in the eastern part of Aphrodite Terra (the planet's largest continent). Human priorities on the Venusian surface are:

1) planting crops to provide food
2) building a water-treatment plant
3) planting different crops to validate terraforming experiments
4) building an airfield to make landings easier
5) exploring future landing sites

Most large cities on Earth are near rivers, and there's no reason to start Venusian civilization off on the wrong foot. Once we also have metal in the area, we will land our crew at the first landing site. The activity of these ground crews will be much simpler than Mars. Venus will already have a breathable atmosphere, with the same air pressure and surface radiation as Earth.

While Venus has much more nitrogen than Earth, it is in its gaseous form (N_2) and is not readily accessible for most living things. On Earth, our nitrogen was made organic (N+)

by cyanobacteria and leguminous plants. Eventually we'll cover Venus with these organisms. On the first day we land, this will not yet have happened, so we'll bring some nitrogen fertilizer and seeds for growing our own crops.

We will bring hundreds of specimens for cyanobacteria and leguminous plants. These organisms are themselves able to turn the abundant N_2 of Venus into the N+ that everything else needs. This discovery was called the agricultural revolution and is the reason farmers rotate their crops. (One rotation of crops is legumes to replenish nitrogen.) This will even be applied on our green stations in orbit. We can use OPA-3s to deliver this fertilizer down to our colony or any other bulky parts we need on the ground.

We will bring worker robots with us to help build infrastructure. We will want an airfield, a fresh-water processing plant, and a barn to store crops. (Yes, I said barn in a space exploration book.)

The first activity will be planting crops for the crew to live on Venus. The crew will still keep scientific data for these plants, but since we're adding nitrogen for them, it will not be a proper terraforming experiment. We will grow some leguminous crops as food without fertilizer, and we'll take notes on the results. The crew will also plant hundreds of different species so they can observe how well they grow.

We will send unmanned probes down to the Venusian surface and make our best guess about what plants will grow the best after the atmosphere and hydrosphere have been altered. We will verify these guesses with the first human landing team. We will not begin large-scale biologic terraforming until their results come back.

Since we picked a site at the confluence of a freshwater source and a salt water source, we can even experiment with aquatic plants, corals, and algae. I say "salt water" to denote water with some dissolved elements. Freshwater on Venus will still be freshwater because the convection cycle that produces freshwater on Earth will be the same as on Venus. The ocean water will be different from Earth's because there will be different things dissolved in it. Earth saltwater species may be fine in Venusian salt water, but we'll have to see. The two planets are made of the same things on a molecular level, so it shouldn't be anything that life won't take to.

The orbits of Venus and Earth align every eight months. The team on Venus will have much shorter tours of duty than their coworkers on Mars. Their main duty will be to keep notes on the biologic experiments and direct the robots building basic infrastructure. We need to know which species grow well under what conditions and for how long. Their findings will be relayed to Space Corporation employees throughout the solar system, who will use this data to finalize the preparations for biologic terraforming on a planetary scale.

CHAPTER 8

Terraforming Mars

Overall, I am the type of person that prefers leaping over chasms to slowly climbing down them. In this chapter, I will deal with terraforming Mars. This plan was developed in parallel with the plan for terraforming Venus. It will be done by three general methods:

1) Photonic terraforming
2) Kinetic terraforming
3) Biologic terraforming

HUMAN-LANDING PREPARATIONS

Previously in my plan, we sent sixteen Katas to the Kuiper Belt as part of the reconnaissance phase. Eight of the Katas will drop a Martian balloon probe into the Martian atmosphere as well as place a communication satellite in Mars orbit. The balloons will radio back precise information, and the satellites will work as an MPS constellation for Mars so we can link the findings of the balloon probes with the specific grid coordinates provided by the satellites. These balloons will be bound to the Martian wind system, but we'll do the best we can to cover most of the planet in two years. From these balloon probes, we'll get data to decide on the first landing site. We'll launch these balloons between cycles of Martian dust storms, and we'll stagger the launches and spread them out to try to cover as much of Mars as we can. The first landing site will be the best place for a human colony, generally near the equator and south of Valles Marineris. This region is about two million square kilometers (the size of Alaska) and will give us plenty of space to pick a specific site.

With all of the data we will have in year nine, the exact position for the first landing site will be selected. The region south of Valles Marineris is ideal for the following reasons:

1) The general location along the Martian equator means it will be closest to Earth-light level.
2) It is as far as possible from the various impact sites.
3) It is a relatively flat area for the surrounding two thousand kilometers, earmarked for human settlement.
4) It is an area that will be near water after the planet floods and gets an ocean. It will not be underwater.
5) The area is buffered by mountains to the south, providing protection from debris from megapili impacts. On the north, there is no elevation buffer, so moisture from the northern ocean will easily come in.

Once the specific site within this area is selected we will divert two pili to the landing site. One will be mostly ice, the other mostly metal. We will make the angle as shallow as possible so that most of the debris will remain together (compared to coming down at a ninety degree angle).

The criteria for picking the specific landing site will be complex. The best thing we could possibly find on the Martian surface would be a site with a marketable quantity of metal. It is expensive sending OPA-2s down to the surface. For cheap common metals like iron, aluminum, copper, and nickel, we'll want a source physically on Mars. Starting out, we can crash a few small pili with these metals on them, but once there is a colony of several thousand people living on Mars, kinetically importing metal will anger our customers. OPA-2s will be needed to import specific elements or to bring in manufactured goods.

Based on the exact position of the debris fields, we may want to adjust the landing site. It would be ideal to park the HAB next to a field of ice boulders and near a large supply of metallic boulders. On many parts of Mars, there is a layer of ice immediately under the dirt. It may not be necessary to send an ice pili into the landing site area, but it will be easy to do. We can also make OPAs for delivering ice and metal.

I have already detailed the first asteroid mining expedition in year eight. Aside from making a large amount of money, these mining expeditions will have two important effects on the Martian landing in year ten:

1) Simple metallic components needed for Mars will be made in space with nonmarketable metal. These components will be placed in a modified OPA-2 and sent down to the landing site. Some of these items will include electric wires, vehicle engines, air filters, mining equipment, power generation equipment, and metal support beams.

2) Nonmarketable material will be used to make infrastructure for propellant and food in space. This will even include propellant and food stations in Martian orbit.

We will not launch manned missions to Mars until we have enough propellant, oxygen, and food to conduct them in a fiscally responsible way.

Landing Site, South of Valles Marinaris

Image 16

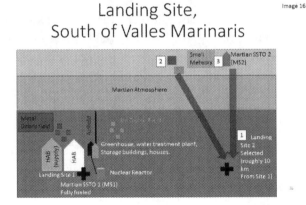

After all of the necessary pili have come down, an MS-1 will land at the site as well. It will land with a fuel mixture of 10 percent extra hydrogen fuel in its tanks. We can mix this hydrogen with CO_2 in the Martian atmosphere to make fuel for the flight out of Mars. We can monitor this from Earth to ensure there are no problems. We'll want the first vehicle to land with its own hydrogen. If an issue does come up, we can send another unmanned MS-1 before the HAB arrives. Once we are confident in this method, MS-1s can land when they are on empty, and we can add the hydrogen through hydrolysis of water on the Martian surface. The MS-1 will fly to Mars unmanned, but the crew compartment will be full of plants and air batteries. It will leave with plenty of CO_2 for the trip. By the time the human crew open it on Mars, it will be full of oxygen. This will allow the crew to have a simple, low-tech, low-energy means of getting oxygen. If something goes wrong with the plants in the MS-1, they can also make oxygen out of water through electrolysis.

We will need two unmanned supply ships to land at the site. All of these movements will occur before the humans land. In case there is a crash or a malfunction, we can send a replacement.

The supply ships will basically be identical to a HAB. The only difference will be that the interior will lack proper sleeping areas, work spaces, and control interfaces. They will be set up for basic human life support and will have minimal communication equipment. Most of the supply ship's contents will be emptied as soon as the crew land, so it will be available as a backup living area. The items on the supply HABs will be as follows:

- two surface rovers (disassembled)
- surface nuclear reactor (modified ESR)
- nuclear fuel

- emergency and supplementary food
- worker robots[10]

The food and fuel for the surface team can be brought in by OPA-2. The more complex components will need a gentler landing and will have to come down in a modified HAB.

We will provide only the core components of complex machines, since we will be manufacturing ceramics and bricks on the Martian surface. We will also have worker robots, which will be able to do simple, automated tasks, though humans will be able to control them remotely.

Once the landing site is expanded after the impacts, these HABs will be empty of supplies and vehicles, and they will make fine living quarters. This would be like converting a very large garage into a two-bedroom apartment.

FIRST LANDING ON MARS

After all of these preparatory steps have been taken, we will actually launch a human crew in a HAB in year ten, with a HAB landing array (HLA-1) attached. The HAB will dock at Green Station 1 and take on three months of food. The crew will replant crops in Green Station 1 before they leave.

After leaving Green Station 1, the crew will attach a large metallic shield made of melted down space debris to the outside of the HAB. This radiation shield will be built to the same specifications as the HAB, only slightly larger. It will lower the ambient level of radiation inside of the HAB and provide some protection against micrometeorite impacts. Once the HAB is underway to Mars, we will keep part of the third-rocket stage from its launch and rotate this section at sufficient speed to give

---

10   NASA is already working on these and calls them Robonauts. We will
     license them separately, so I'm calling them worker robots.

the crew artificial gravity. Technically, the force we will be using is centripetal force, but the crew will think of it like gravity. They will be very bored and spend most of their three months watching movies or reading Wikipedia, until they conduct a gravitational breaking maneuver near Mars.

In Mars orbit, the HAB will dock with the Mars green station. The crew will change out the air batteries, collect three months of crops, and exchange any other waste they have accumulated. It may not be practical to replant crops now, since no one will visit this station for two years. We'll try to coordinate this with the space tourism business. Hopefully every 3 months a 'solar system cruise' can dock with Mars green station (some people would pay money to work in our space-farm for a day). This will keep the seeds and crops in rotation as well as make some money on the side. Whenever it's time to come home we'll ensure no tourists drop by that month. We'll also leave a worker robot on Mars Green Station 1 that can be controlled from the Martian surface. If no space cruise ships come by the robot can replant the crop from seeds left on the station.

After leaving the Mars green station, the metal shield will be removed from the HAB and placed in Martian orbit. The crew will proceed to land on Mars. NASA has placed many large space probes on the Martian surface; we'll adopt their general method for landing. They use an aero brake to slow down in the upper Martian atmosphere, a parachute to slow down for most of the way down, and finally a small rocket at low altitude to bring the object in for a soft landing.

ACTIVITY ON MARTIAN SURFACE

As soon as the crew land, they will unload a nuclear reactor designed for use on the Martian surface, as well as remote-controlled worker robots.

The landing team's priorities on the Martian surface are:

1) building greenhouses for food-crops
2) building several greenhouses for researching different crops
3) building a water-treatment plant (connected to greenhouses)
4) building an airfield for subsequent landings
5) exploring future landing sites
6) primitive road building

The reactor will provide the heat and energy to make bricks and glass. They will also use this technology to build an airfield at their landing site. It will be much cheaper for all of the subsequent teams to land on Mars in a horizontal configuration instead of a more expensive vertical one. This is why airplanes on Earth land on a runway and not with an aero brake, parachute, and booster rocket.

Since we had the forethought to land a metal pili and an ice pili in that general area, it will be easy to mix water with Martian soil and apply a little heat from our nuclear reactor to make brick. For better accuracy we can use an OPA of steel or ice incased in steel, whatever we need. To make glass will be more complex, but we should be able to separate silicate-rich soil on Mars and heat it enough for glass panels.

We will have ultraviolet lights on hand to add light to crops if needed, but Mars has a comparable day-night cycle to Earth, and we'll be on the Martian equator. We can plan for the plants to use natural light. (During a two-month dust storm, we may need the ultraviolet lights.) We will also bring simple air filtration systems so that the plants can live off of CO_2 from the Martian atmosphere, without depressurization. We will distribute the oxygen and CO_2 between the HAB and the greenhouse using air batteries at first. With time we'll have proper air hoses and pumps connected across the landing site.

Once the greenhouse is built, we may need to spray the soil with water and a few other elements to make it more suitable for terrestrial plants. After that, we'll plant all of the seeds. The crew will of course have normal food actually brought from Earth and waiting in Martian orbit. At no point will they have a risk of starvation. Backup food can be viewed as a delicacy. During a two-year surface stay, a four-person crew will literally consume tons of food and oxygen, and it will be much cheaper to develop this capability on Mars. I want the same container of backup food to stand untouched on Mars for a thousand years.

After the Earth-standard greenhouse is built and the crops are planted, we will begin several different activities. We will build a few extra Earth-standard greenhouses, since there's no reason to be cheap about food. We are planning for the long term, and these greenhouses can be turned into useful buildings after the terraforming event.

We will build several different types of experimental greenhouses. We'll find a small crater on Mars and dome it over to create an artificial pond. We'll also want to make a separate greenhouse and bring mammals there to test the long-term effects of low gravity. While the human crew will be very healthy adults, I want actual mammals born and grown on Mars to see what the effects of low-gravity would be on animal growth. Bringing mice, small pigs or goats to Mars would be very useful to study gravitational affects. Pigs and goats can also eat waste products from our farm on Mars (this is why farmers on earth have them).

For natural resources, we will keep our eye out for minable quantities of nuclear fuel and cheap metals. It is not cost-effective to bring iron, copper, and nickel down to Mars by OPA-2 on the scale that Martian civilization will require. Exploratory work on locating these assets will do much for increasing settler interest on Mars.

We will bring air-filtration technology with us to collect the CO_2 from the Martian atmosphere under low pressure and make it consumable by our plants at normal Earth pressure. While we are doing this, we can filter the air for hydrogen deuterium (H2D). H2D is five times more abundant on Mars than on Earth. Hydrogen deuterium is used in nuclear power plants. It is valuable enough that we may be able to export it from Mars to Earth. (It's worth US\$10,000 per kilogram.) At the very least, we can store it on Mars so that it can one day be used for the Martian nuclear industry. We will never be able to use fossil fuels on Mars (no fossils). From the very beginning, Mars will be an atomically or solar-powered society so the H2D will not go to waste. Who said you can't make money out of thin air?

The presence of nuclear fuel on Mars will be of immense value to our human settlement—to such an extent that if we find a good supply on Mars, we may want to plant a colony next to it. A useful geothermal vent or the likelihood of good soil for agriculture once the climate is changed would also be a consideration. In the long term we will use solar and wind power for our settlement of Mars. The nuclear power is what will build the solar and wind power infrastructure.

While I want to have a surface team for general exploration and the inevitable things that computer programmers and engineers can't foresee, we can supplement the landing team with space-based human operators if needed. When I was in the US army in Afghanistan, many of our Predator drones were operated from Nevada, more than twelve thousand kilometers away. To have humans operate ground robots on Mars from Martian-synchronous orbit would involve a radio transmission from seventeen thousand kilometers. Far from being futuristic, this is decades old. While space-based Martian robot operators won't be able to impress girls in bars as much as the astronauts that land on the planet, it will be much cheaper to keep extra personnel outside of the Martian gravity well. Besides, lying to girls in bars about being an astronaut won't be as effective when there are a hundred thousand astronauts.

While we can operate these robots remotely, we still can't necessarily operate them from Earth. Radio signals travel at the speed of light. Inside of the moon's orbit, this is a delay of about one second. The time lag between Mars and Earth is an average of 20 hours. That's only one way. For you to toggle a control to move a robot on Mars, you would not observe the effect of that instruction for an average of 40 hours (talk about slow internet!)

EXPERIMENTAL GREENHOUSES

The first thing each crew will do after they set up their nuclear reactor is to build a large greenhouse and plant crops. After this, they will start building experimental greenhouses. We will, at a minimum, have four different types:

1) Space Corporations' best guess on the post-impact Martian atmosphere, moisture, and light level.
2) The best guess for the Martian hydrosphere (even if we have to build a small pond). This will be vital to test cyanobacteria, plankton, fish, and seaweed. If possible, we can dome in a small crater to use as a lake.
3) A very large greenhouse to test full trees in Martian soil, under simulated conditions. We'll even have to heat that part of the crust down several meters. (Trees can be as deep as they are tall.)
4) A large greenhouse to test a group of small mammals. Pigs and goats would be good both for science and agriculture (they can eat parts of plants that we can't). We may wait a few years until we have enough food and

domed in space before we bring a breeding population of animals to Mars.

We will plan to have several different landing sites, but all within 10-20 km of each other. The goal is that a problem with a nuclear reactor, or a crash while bringing in a new crew would still leave us nearby reserve facilities that are working fine. Once the colony expands after the impact event the space between landing sites will be full of crops and forests. If we find very useful elements we may space them out more (aluminum mine, uranium mine, geothermal vent for example).

Whenever we land, the landing site will be prepared with a ten-thousand ton ice pili and a ten-thousand-ton metallic pili. The Martian atmosphere is very thin and impact velocity on Mars is much lower than it is on earth. If we bring the megapili in at a shallow trajectory we can mitigate how much melts during atmospheric entry, or how much is thrown out after impact. These impacts will ensure that when the HAB comes down it will have plenty of material that our landing team needs to build infrastructure without scouring the planet to find it. We can also use drag shoots to slow the descent of raw materials without the expense of a vehicular landing.

We can also bring in small amounts of ice or metal using OPA-2s. We could bring 100,000 tons or a million tons down by pili, OPA-2s are usually 200 kg. To make this more practical for our timetable, we can place dozens of asteroids of this type and size in Martian orbit with radio tags attached. Once we decide where the landing sites are going to be we can send in the pili, even multiple pili per site.

After one year and nine months, Crew 1 will control a robot on board the Mars green station and plant crops for their journey home. Seeds can be stored for very long periods of time, and the crew will have left their packets of seeds on the station before they went down to the surface. Three months later, Crew 1 will remove all of the plants from the MS-1, put them into a greenhouse, board their MS-1, and fly to Martian orbit. In Mars orbit, there will already be a HAB with an attached radiation shield and attached ballista docked at the Mars green station. Crew 1 will transfer all of the air, food, and water they need for the journey back to Earth into their HAB. The MS-1 will be refueled with hydrogen and left in Mars orbit until Landing Site 3 is picked. (The MS-1 for Crew 2 will have landed a year prior.) Crew 1 will proceed to Earth in the HAB. In Earth orbit, they will leave the HAB and board an ES-1 for the ride back to Earth's surface. Besides this being the most significant event of the third millennia, the price of Space Corporation stock will be out of this world.

Crew 2 will land just before Crew 1 leaves the surface. Their mission will be identical to Crew 1, except Crew 2 will land ten kilometers away and will draw their first load of food from the crops Crew 1 planted. The two-year window of proximity between Mars and Earth is not open long, so only a brief handover will be possible. The crews will have plenty of time to speak and share notes by radio during their space flights in the vicinity of Mars. An actual linkup on the surface will be dependent on many factors, but it will be possible.

Each crew will have the ability to drive to any other site if they need engine parts, surface suits, or extra air batteries. While we will need a continuous presence on the Martian surface, the robots from each site can be controlled by humans at the adjacent facilities, even from orbit.

These landings will continue at two-year intervals until year forty. The colonization zone (the two million square kilometers south of the Valles Marineris) is the area farthest from all the planned impact points and is buffered by mountains between it and the nearest impacts. At most, there will be a risk of steam, dust, and water breaking pressure domes or burying equipment and buildings. Since both the impactors and the locations they target will be made of lightweight volatile material, there should be relatively little solid debris thrown up from the impact event. Nonetheless, we will evacuate all humans and prepare all buildings and equipment for light damage.

GREENHOUSE GAS PRODUCTION ON MARS

Dr. Zubrin has estimated that thirty-nine million metric tons of chloro-fluoro-carbons would be enough to terraform Mars.[11] I am substituting methane from the outer belt for CFCs. Methane has a greenhouse gas factor of seventy-five, and CFC-12 has a greenhouse gas factor of eleven thousand. I will multiply the thirty-nine-million-ton figure by 152, which results in six billion metric tons. The four Kuiper Belt objects selected by the Katas will be picked primarily for methane content. One ice object that is four kilometers in diameter will be about fifty-six billion metric tons. Even assuming a mixture of water-ice and methane we should easily be able to get six billion metric tons of methane. Methane does break down after about 20 years. In the long term we can plan for Martian industry to produce CFCs as a by-product of road and airfield building on the Martian surface. Ironically, CFCs were created as a by-product of manufacturing here on earth. They were so common in industry that, far from paying people to create them, the government had to ban their manufacture by law. On Mars we will simply encourage their manufacture (deduct the cost from the 10% funding rate on space commerce) until the atmosphere has the correct greenhouse factor. While CFCs are far more efficient at increasing temperature, they are impossible to find and difficult to make off of Earth. There is more than

11   See http://www.users.globalnet.co.uk/~mfogg/zubrin.htm.

enough Methane floating around in the Solar System right now. Chucking four giant methane snowballs into Mars is elegantly simplistic.

Methane also has a very low melting point. There is virtually no methane found in the main asteroid belt, because it would have melted under the relatively warm conditions in that part of the solar system. If there is any at all, it has only persisted because it was blocked from direct sunlight by water-ice or rock. Water-ice has a higher melting point and breaks down slowly in the main belt. As frozen methane objects approach the orbit of Saturn, we will cover them with thin aluminum reflectors. These will keep them out of direct sunlight during their trip to Mars. We'll remove the ARs just prior to Martian impact. Below is a table of different greenhouse gases and how good they are at retaining heat.

Table 6. Greenhouse gases and their warming factors

Gas name	Chemical formula	Lifetime (years)	Global warming potential (GWP) for given time horizon		
			20-yr	100-yr	500-yr
Carbon dioxide	CO_2	Forever	1	1	1
Methane	CH_4	12	72	25	7.6
Nitrous oxide	N_2O	114	289	298	153
CFC-12	CCl_2F_2	100	11,000	10,900	5,200
HCFC-22	$CHClF_2$	12	5,160	1,810	549
Tetrafluoromethane	CF_4	50,000	5,210	7,390	11,200
Hexafluoroethane	$C2F_6$	10,000	8,630	12,200	18,200
Sulfur hexafluoride	SF_6	3,200	16,300	22,800	32,600
Nitrogen trifluoride	NF_3	740	12,300	17,200	20,700

CFC-12 produces almost 152 times more heat than Methane. We can easily move billions of tons of methane from the Kuiper belt to Mars. We can also integrate a small amount of CFC production as a by-product of manufacturing tools, bricks, and ceramics on the Martian surface.

Most engines on Earth take in gas from the atmosphere, mix it with other substances, and then emit some gases as waste. We can design our equipment on the Martian surface to take in CO_2, harvest H_2D, and emit CFC-12. By the methods I have outlined in this book, we could not produce thirty-nine million tons of CFCs unless we were able to make about eight billion tons of bricks. In the long term, we will make 8 billion tons of bricks (the first small town on Mars will weigh much more than that).

Chlorine is not common on Mars; at most it is 0.5 percent of the Martian crust by mass (probably much less). So to use chlorine to make millions of tons of material means that we have to heat that much Martian rock, capture the chlorine that comes out, and combine it with fluorine (equally as rare) to make exotic gases. Every little bit helps, but the greenhouse gases on a planetary scale will not come from the planet we're greenhousing.

ECONOMIC ACTIVITY ON MARS

I appreciate that any human activity in space that has the express purpose of generating revenue is completely contrary to everything we've come to expect from space exploration. Most exploration on Earth was done in expectation of profit. There is virtually zero scientific exploration of the earth's crust; it is primarily for-profit oil companies. Even the Chixulub crater in Mexico was found by an oil company, not scientists poking around the earth's crust for fun. Vasco da Gama, though he lost many of his men and ships during his journey from Portugal to India, was able to make a 6,000 percent return to the stockholders that financed his voyage. In what was certainly a coincidence, a huge number of Portuguese became involved in the for-profit colonization effort the moment they heard about this.

Keeping costs down and profits up is the only way that Mars and Venus can be colonized. The primary purpose of landing a human team on Mars is to conduct experiments toward biologic terraforming. But the more developed the colony is, the faster we can settle Mars after the impact event.

Many of these projects can be finished during the first two-year landing. After that, the subsequent crews can expand on the earlier projects or make a landing site a few kilometers away. It would be great to have three fully functional sites with separate infrastructure for each one. This would be especially good if one had a uranium mine, one a geothermal vent, and the third an aluminum mine. If we want to do extra work on the airfield at Site 2 or if we have a problem with the water filtration system at Site 3, the other sites will be backups. Once colonizing Mars begins in earnest, all three of the sites will be fully occupied with hundreds of colonists.

All of the buildings on Mars will be made from bricks and ceramics of Martian mud, and made mostly by worker robots controlled by the crew. Any complex elements or machines we need can be brought in by OPA-2s. Since we have to bring this capability to Mars, we can keep these largely autonomous robots working while the humans supervise the robots and conduct biologic terraforming experiments.

We will design any worker robots that we use on Mars so that they can eventually be fitted with seed-planting equipment and used as primitive tractors after the climate has been changed.

This will not occur for thirty years after landing, but robots will be kept adaptable and multipurpose in our plan.

Each human landing site will be powered by a nuclear reactor. This will also provide the heat for making bricks, ceramics, or glass. The humans will only handle irradiated material by remote control, using a robot. As each reactor is set up on the surface, the humans and worker robots will cover it on all sides with dirt and mud. This will buffer the humans from radiation.

The reactors could even be positioned one kilometer away from where the humans live and on the other side of a hill or terrain feature. Power and data lines can go out to the reactor at a considerable distance, and a small backup generator will be available on the HAB itself. In this way, even the small chance of a nuclear reactor problem would simply entail visiting a neighboring site while your robot fixes the reactor.

Since the core concept behind Space Corporation is to profitably go to space, this will be factored in to our choice of landing site. On Earth, nuclear fuel is found in only a few areas, but it is found. It goes to reason that Mars would have a few very rich veins of precious or useful metals. Kalgoorlie, Australia, has produced a huge amount of gold, and Macarthur River in Canada has produced a huge amount of uranium. Since we own all of Mars, we can use the data from the balloon probes to pick the best site for mining. We can find the best spot to mine uranium and put our landing site there. We may not export uranium from Mars to Earth, but we could certainly use it on Mars.

Just like any large corporation on Earth, we will set up subsidiary companies of Space Corporation for Venus, Mars, the asteroid belt, and so forth. In this way, we can keep accountability of money, work, equipment, and personnel.

As Martian society grows, it will be nuclear and solar powered from its infancy. Without fossil fuels, everything on Mars will be run by nuclear reactors, and planting Martian society on a uranium mine would be the best thing we could do. Which specific metals are imported to Mars versus produced on Mars will be decided by the free market.

KINETIC TERRAFORMING OF MARS

The orbit of Mars around the sun is more elliptical than Earth's. Once every 686 Earth days, Mars reaches perihelion (closest point to the Sun). Earth is roughly the same distance from the Sun all year, but Mars is 10% closer to the sun on its perihelion, warming the planet considerably. This causes the atmosphere to be hotter and denser than any time during a Martian year. The hotter the atmosphere is the easier it will be to terraform the planet. Even the LaGrange objects will be able to bounce more light to the planet, since the distance their light is traveling will be at its lowest point. We will plan phase I of kinetic terraforming to occur on perihelion 40 years after Space Corporation is founded.

A core idea behind kinetic terraforming is impactor optimization. Quintillions of golf ball sized pieces will melt before reaching the surface of a planet, but this would be cosmically expensive (stars by then will be cheap). While it would be very simple to divert the 200 km Chiron into Mars intact this would cause massive damage to the planet (crust destroyed, mantle disrupted, inner solar system debris, etc). The more the crust is disturbed the more energy is wasted, dust kicked up, geographic features needlessly created or destroyed. Areas with kilometers of Solid ice will be targeted kinetically. The enormous heat from these impacts will melt the smaller concentrations of ice that are too thin to impact with a megapili.

It will take a small portion of these impactors to heat the thin layer of CO_2 covering the northern and southern caps. As the atmosphere gets thicker and hotter this will increase the temperature of objects entering the Martian atmosphere. The first batch of impactors will be set to ballistic cratering size plus 50 meters (to account for part of the object melting in the Martian atmosphere). The second batch can arrive a few minutes later but they will find a very different atmosphere to fly through. Batch 2 will be ballistic cratering size plus 60 meters. Batch 2 will heat and thicken the atmosphere, so batch 3 must be 70 meters larger than its ballistic size. When we have finished breaking Okyrhoe apart and bringing each piece down in its specific location the atmospheric friction allotment will be closer to 250 meters per impactor (wave 1 will be more expensive than all of the other waves, even though it's the least mass. Fusion devices are cheap per yield. Cutting megapili into small specific pieces is expensive. The larger the pieces, the fewer blast seams we need to make, the less operational cost.

Even a slight increase in size saves us money in cut out megapili (larger means fewer objects, less cuts, less energy, less man hours, less maintenance). A reasonable plan is to create 10 different batches of impactors (both north and south). Each batch will come down in a grid pattern over its pole. Each impactor will be custom-sized to the depth of ice in its assigned impact circle.

There are many different resources to make projections about impactor size, mass, velocity, and crater depth. I have links to the site I used for this portion of my book below.[12] The numbers below do not factor icey objects partially melting during atmospheric entry.

---

12   http://pirlwww.lpl.arizona.edu/~jmelosh/crater_c.cgi

North Polar Cap: (1100 km in diameter, average depth 2 km) 949,850 square km

Standard Northern Impactor (SNI): 1400 m in diameter
Crater diameter from SNI: 8.2 km (53 square km).
949850 / 53 = 17,921 Impactors needed to crater the North Polar Cap.

Diameter if all impactors come from the same object: 36 km.

Crater depth 2 km
Assuming 45° angle for impact and velocity of 7 km/s (slowest possible for Martian gravity).

South Polar Cap: 400 km diameter, average depth 3 km 125,600 square km

Standard Southern Impactor (SSI): 2200 m in diameter

Crater diameter from SSI: 12 km (116 square km) Crater depth: 3 km
I'm assuming a 45° angle for impact and velocity of 7 km/s.

125,600 / 116 = 1082 impactors

Diameter if all SSI come from the same object: 22 km.
For this plan both sets of impactors will come from the same comet: 52872 Okyrhoe.

The Main Belt comet 52872 Okyrhoe is 49 km in diameter. We know that Okyrhoe contains some volatile materials. Of course we will make specific decisions on which asteroids to use after the Katas have conducted their survey, but Okyrhoe is the right size for my purposes and I'm using it as my example.

We will atomically mine Okyrhoe in order to adjust its orbit. By cutting one megapili out and ejecting it with a fusion device we can push it sunward into a Mars intercept. We will have ballista-2s on hand to make specific course adjustments. It will take ten to fifteen years of coasting and minor adjustments before Okyrhoe intercepts Mars. During this interval we will be using Ice-Blade Rigs to cut Okyrhoe into 17,921 Northern impactors and 1,082 Southern Impactors. As more and more impactors intercept Mars the atmosphere will warm and thicken. The first objects that impact Mars will melt a little on the way down, so we need to make them a little larger (1450 m and 2230 m for example). Since thousands of impactors turning the poles to steam will thicken the atmosphere we'll have to design the next wave to be slightly larger (1500 m and 2260 m). Okyrhoe is big enough to provide all of these impactors with plenty to spare for melting on the way down. There is also a few hundred meters of ice spread all over the bottom of the Hellas Basin. Before the large impacts we can send a few thousand small objects down to Hellas Basin to break up the small concentration of ice. After the atmosphere is turned to thick steam it will be harder to accurately aim small objects from orbit to the surface. The goal is to melt/crater all of the CO_2 and water ice on the Martian surface without unnecessarily digging up dust. A lot of dust will be brought into the atmosphere anyway, but we will minimize this by not targeting dust.

We will use the balloon probes to get precise data on both polar caps, and any other ice concentrations on Mars. I've planned the use of 8 balloons as part of our initial reconnaissance. We can certainly send another wave of balloons before the impact event if we need to. The calculations on crater depth and impactor size are pretty straightforward. We can divide the North Pole into

17,921 impact circles, give each one a unique designation and do precise math on each impactor.

If impact circle #7432 is 2.6 km deep we can make megapili #7432 exactly 1600 m in diameter (ballistic size). If we assign megapili 7432 to the third set of impactors we will know that 8 km of steam will be covering Mars at the time it lands. We will cut it 80m wider then ballistic size. It will be 1680m wide when it enters the Martian atmosphere and 1600 m wide when it touches the Martian surface. We'll plan all 19,000 impactors to arrive in 10 sets so we can reasonably factor in changes in steam and air pressure between each set. This might sound complicated but Field Artillery units in the US Army do much more difficult math then this when plotting a fire mission.

It will be expensive and tedious to break Okyrhoe up in this way, but it will allow us to give Mars a temporary super-heated steam atmosphere. As the pieces are cut free ballista-2s will be used to place them on a precise impact course for that specific impact circle on Mars. We want these impacts to take place at roughly the same time, but they certainly don't have to be simultaneous. I think a 4 day window is very reasonable. We can accelerate each object just enough so that they are in line with 5-10 km between each one. Mars is also moving of course so the line would be slanted.

Each megapili will be moving at 7 km/s or slightly faster. We will synchronize this with perihelion on Mars (the hottest day of the Martian year) so that everything melts better. All activity will take 4 days, with the first impactors arriving on Perihelion. After Okyrhoe has come down on the Martian surface all of the CO_2 and water already on Mars will melt and turn into a huge steam atmosphere 40 km thick. Okyrhoe's evaporated nitrogen will add about .5 km of superheated gas to this atmosphere as well. There will also be a large amount of trace gases that are warmed up from the soil, as well as frozen water accumulated on Mars that was debris from megapili sent to Venus.

KINETIC TERRAFORMING PHASE II:

Okyrhoe is only a small portion of the mass Mars will need to change its atmosphere. Most of the mass will come from 3 sources: 8405 Asbolus (84 km in diameter) 174 P Echeclus (84 km in diameter) and 2064 Chiron (240 km in diameter). We will be breaking these comets into smaller pieces to melt them properly, but we will not use the same method we did for Okyrhoe. Now that the atmosphere has been super-heated, an ice object of 500 m diameter would completely melt in the Martian atmosphere. We can comfortably make 600 m diameter pieces, since a few chunks of ice reaching the surface will not upset our plans.

All 3 of these objects will be diverted by atomic mining and ballista-2s, setting them on a Mars intercept. Chiron is very far away (near the orbit of Saturn when it is at perihelion). The best way to send Chiron to Mars is to alter its orbit so that we can have a gravity assist with Saturn or Jupiter. Over the span of 40 years we will have enough time to do this, even if we have to start adjusting the orbits 25 years before the impact event.

We will use ice-blade rigs to neatly cut Chiron into 8 roughly spherical or cubical pieces the size of 8405 Asbolus and 174P Echeclus (80 km in diameter). For simplicity I'm naming these Chiron 1-8. We will then use a mass produced set of Mtesrs to perforate all Chiron pieces, Asbolus and Echeclus with holes radiating out from the center of mass to the surface. Since all of these objects will be 80-90 km in diameter we can make a set of Mtesrs that are 50 km long (so they can melt from the surface to the center of mass on all sides). We can use the same set of equipment on all of these objects and save equipment costs (we'll have years to cut these tunnels). We'll melt from the surface down to the center of Mass, and move around the surface as needed. Eventually all 10 of these objects will be perforated with blast tunnels. We will use a TESR to make a large tunnel down to the center of mass and place a large fusion device there. We will use the Mtesr to make small tunnels that are 600 m apart (in the same way that paper letters have a portion that is artificially weakened so it is easier to cut or tear by hand). When the Fusion device in the center detonates it will send super-heated steam through millions of Mtesr tunnels, resulting in millions of 600 m ice pieces.

We can control the yield of the device so this energy will be enough to break the object into many pieces, but not enough to propel the pieces any reasonable distance away. Since the object will already be near Mars and have the momentum of moving forward all of the pieces will coast toward Mars, spreading out enough to come down all over the planet. Mars has 144 million square miles of surface area. The pieces from all of the phase 2 impactors will be spread out over the surface of Mars and separated by several hours. Chiron 2 can be cut into slightly larger pieces since Chiron 1 will have heated and thickened the Martian atmosphere.

Everything will be the same except this piece will be several hours behind the first phase 2 impactor. Over the course of ten days all of these objects will have come down into the super-heated steam atmosphere, adding more and more nitrogen, water and heat. I am assuming that these objects are primarily made out of nitrogen/ammonia (ammonia is 80% nitrogen by mass). After the Reconnaissance phase we can validate or invalidate this assumption. We may add more objects, or use completely different ones. When all of these pieces are added Mars will have 4 E 18 kg of nitrogen (the same amount of nitrogen that earth does). The more water the better but we can add water later if needed. Some heat will be radiating out into space, but even more heat will be generated by all of these impactors. Since total energy is conserved friction from a 500 m impactor melting up in the atmosphere is the same amount released if that objects were to crash into a solid object and produce an explosion. The atmosphere of Mars will now be 100 km thick and extremely hot. This will set us up for the next phase.

KINETIC TERRAFORMING PHASE III:

Phase III of kinetic terraforming will be very similar to phase 2 except that the content of the impactors will be different. We will divert asteroid 596 Scheila, and the Jovian Moon Amalthea. If we have to we can use Saturn's moon Iapetus (which has enough water for all of the crazy schemes in this book combined). As responsible business people we will use the closest possible asteroid to keep costs low.

The atomic mining method works for moons as well as asteroids and comets. The asteroid 596 Scheila and the Jovian moon Amalthea will be directed to Mars. They will also have Mtesr tunnels added. Once they get near enough to Mars we will detonate a fusion device in their center of mass and millions of water pieces will enter the Martian atmosphere at all points. Water starts to go through thermolysis at 2000°C (though only 3% will react at this temperature). At 3000°C 50% of water will go through thermolysis. At 6000°C 100% of water goes through thermolysis. This was the same effect we used to terraform Venus in Chapter 7. Each of these impactors will release a huge amount of energy, as long as we can concentrate all of these impactors in a relatively short period of time, the heat will not be able to dissipate. As these water impactors come down on Mars the water will split into hydrogen and oxygen.

The oxygen will be the heaviest element in the Martian atmosphere, and will sink to the surface. The hydrogen will combine with other elements or be blown into space. At this point the Martian atmosphere will have the same composition as Earth's atmosphere (except for being phenomenally hot, and having an ocean worth of steam floating around). We will ensure that there is sufficient CO_2 in the Martian atmosphere to keep the planet at Earth norm. It will be slightly higher than earth (Mars needs more greenhouse gas since it's farther from the sun). We can factor in a few methane impactors from the Kuiper Belt to get the total greenhouse gas level just right. We will of course keep the toxicity of our greenhouse gases in mind when terraforming Mars. Since Venus will have the same sunlight as earth we'll ensure it has the same proportion of CO_2.

Methane has no harmful effects on humans and has a very high heat factor relative to CO_2. Even a few billion tons of Methane from the Kuiper belt can help the Greenhouse effect on Mars. All of the heat on Mars will start radiating out into space. With no additional impacts to keep the temperature high Mars will eventually go from thousands of degrees Celsius down to Earth norm. As the planet cools the steam in the atmosphere will condense into clouds. As the clouds cool this huge amount of water will fall down on the Martian surface as rain. This will help cool the planet's surface, as well as wash most of the remaining dust from the air. Dust is carried by hot hair, so as the planet cools the dust will come down to the surface as well. We may have to wait a few months for the planet to cool but when we're done we'll have an Earth standard atmosphere. We can also divert the LaGrange objects while Mars is cooling down so the 5% extra sunlight will not be present. Once the temperature gets to earth norm we'll realign the LaGrange objects to bring the sunlight level back up.

MEGA-CHEAP CANALS

The Mars impacts in year forty will be timed to occur during Martian south polar summer and perihelion, when the atmosphere is ten times denser than usual. Besides general terraforming, we will have some specific geographical adjustments to make. While terraforming will mean changing the global climate, it also means changing the shape of that globe. We will be building a three-hundred-kilometer-long canal from the Hellas basin to Vastitas Borealis with megapili. We will use leftover pieces of Okyrhoe.

Currently Martian topography has a huge basin in the northern hemisphere, Vastitas Borealis, and two large inland basins, Argyre and Hellas. Argyre is three tiems the size of the Mediterranean and Hellas is five times. There is no liquid water on Mars now, but the lowlands are the places that will fill with water first. Argyre is about one hundred kilometers away from Vastitas Borealis, across relatively low-lying land. If the planetary water level is high enough, Argyre may be connected to the Martian ocean. The Hellas Basin is about three hundred kilometers from Vastitas Borealis and has a relatively high area of land between it and the future ocean. Being formed from a large asteroid impact, it is a huge circle. It's also right next to the equator, will have very calm waves, and will generally be a great spot for widespread real estate development. Economically and biologically we need to link the Hellas Basin to the Northern Ocean. One can imagine the history of the Mediterranean if it were blocked from the Atlantic. I don't want some Martian colonist to have to figure this out in 200 years; we can do it in ten minutes during the impact event.

Hellas is immediately next to the equator. It is a perfect spot for humans to have access to the sea for trade, fishing, beaches, and generally high-property values. Most humans on Earth live within 150 kilometers of the sea, so we should assume the same for any planet where humans would wish to live. Fortunately, Venus is much more suitable for this than Mars, but I'll address that later.

Aside from helping the circulation of water and air on Mars, connecting the Hellas Basin to Vastitas Borealis will also make terraforming easier. As we put life into the Martian sea, that life will be able to spread throughout the large inland seas—we won't have to terraform each lake. We will ensure that ten of our megapili come down along the closest line between the Hellas Basin and Vastitas Borealis. If the megapili come in at a low trajectory, the craters will look more like trenches and less like bowls in the soil. We will plan for the craters to be deep enough to allow the ocean through once the planet is terraformed. Not only will this be much cheaper to build than the Panama Canal or the Suez Canal, but it will be more fun to watch!

A little forethought when we are designing our worlds will have considerable payoff later. On Earth, doing this would certainly result in a noise complaint. Since we're doing this before anyone has bought plots on Mars, this will simply be landscaping.

I'm planning only one megapili canal north of the Hellas Basin. There are several other lakes in the area of the Argyre basin that could be connected to the ocean by canal, but this will depend on the exact height of sea level we want on Mars. This basic method of creating a canal can be employed anywhere on Mars in order to connect all of the low-lying areas to the sea. Real estate developers hate oceans, and they hate deserts, but they love beaches!

MARTIAN BIOLOGIC TERRAFORMING

Our scientists will spend a lot of time before the impact event conducting studies of different plant, algae, plankton, and bacteria species under conditions we estimate to be like post-impact Mars. The best way to study the effect of Martian light, gravity, and soil on Earth plants is to use Martian light, gravity, and soil. We will alter the atmosphere used in the greenhouses to be the same as Mars after all of the impacts are complete. The ground crew will do this for thirty years (from year ten to year forty). Everyone will leave the surface for the actual impact.

After the impact, we will send down another human landing party. We will have to watch this closely; now that the atmosphere is thicker, we will have to modify our landing equipment (different-size parachutes and rockets). Airfields should be intact after the impact event, though covered in dirt and dust.

Plants on earth have co-evolved with bacteria in the soil that roots need to break down nutrients and thrive. We can cover all of the seeds with a thin layer of dry dust holding these bacteria spores. Once the seeds land in warm moist soil the bacteria will reproduce and spread out. By coating the seeds in this way we can ensure that we have the right bacteria next to the seeds. We will use a handful of these canisters and study the results before we cover the planet with them. Seed canisters will be made from cheap metal that is a by-product of mining. In particular, silicon has a very high melting point, and we will be refining silicon anyway to make the solar panels on the VL1 object. We will already have the capacity to make seed canisters, since we will be using them on Venus first. We will either use identical seed canister, or at most very similar ones. We will have to adjust the specific design to accommodate the difference in gravity, but since we are waiting until after kinetic terraforming in both cases, the change on the seed canister-2s will not be great. I'm assuming that the same species that we use on Venus will also grow on Mars, hence the same seeds. In both worlds, there is a shortage of organic nitrogen, so we will start primarily with leguminous plants on land and Cyanobacteria in the sea.

Once the temperature goes down to terrestrial levels we will land a human crew on the surface to verify or modify our previous experiments. Based on these readings, we will tweak the design for the seed canisters and then test a few of them near the human landing site. If both the canisters and the seeds survive the journey and begin to grow on the surface, we will proceed to send thousands of seed canisters with trillions of seeds all over the planet.

The seed canisters used to breach the Martian atmosphere will be mass-produced, but the mixture of seeds used will vary considerably. Of course sunlight, moisture, and soil composition will be ascertained by sensor data before the entire planet is seeded. An example of this is using tipuana tipu trees, peanuts, and soy along the equatorial region of Mars, planting alfalfa and clover in the middle latitudes, and then using only moss and lichen in the extreme north and south of Mars.

We can specifically designate certain areas that will be planted with clover and alfalfa throughout the Martian surface so that we can use these as future landing sites. We can make a reasonable estimate of the first one hundred landing sites/future cities and then plan a road system connecting all of them. We will not build any of this during the first forty years but we can certainly seed smaller plants along future sites that we want to build on. This will give us easy driving and landing sites on top of smaller plants (like grass on a football field). While most of Mars will be forested we will not have to clear cut a forest to build our roads, airports, seaports, and space ports. Other than these patches and planted areas near the first three landings sites, the whole planet will be covered with seeds from orbit.

In particular, a great place for a city outside of the colonization zone would be where the Hellas canal connects into Vastitas Borealis. We'll want a small settlement at the foot of Olympus Mons for people to go skiing (this will definitely be a stop when solar system cruise lines become popular). Incidentally if we ever made a rail gun to throw freight into orbit from Mars, Olympus Mons is the best place to build it. We'll want a landing site near Valles Marineris for the brave to rappel down and the cautious to fly over. Any deposits of useful metals will be earmarked for future landing sites and of course planned to be near the future coast of Vastitas Borealis (like the Pacific Coast Highway in California).

Other than a few test canisters, the immediate area of the colonization zone will be organized like a farm on Earth. We will plant food crops, animal feed crops, and whatever else we think our growing colony will need. Trees grown in a grid formation grow bigger and taller than trees that grow naturally, and they make a better source of timber. For all of the futuristic 1950's technology that we are using, we must remember that the goal is to create a place where people from Earth will want to live. Most homes, furniture, and countless consumer goods are made from wood, not exotic space metal.

The mega-green stations will be a cheap source of seeds for these various species already in space. If it's cost effective we can fly seeds up from Earth to accelerate biologic terraforming. Since many plant species produce tens of thousands of seeds a year our megagreen stations in space may be adequate. The faster Mars is terraformed, the faster and more it can be sold for. Even if seeds had to be flown up from earth we would bring them down on Mars in seed containers made in space. We may have to genetically modify them or at least selectively breed them.

Mars doesn't have insects to spread seeds around, so we will choose plant species that spread seeds by wind and are capable of asexual reproduction. The first generation of seeds will be intentionally spread out very thin all over the planet. Not only is it easier to spread these seeds from aircraft or from orbit, but a thin distribution of these plants will ensure that the thousands of seeds each plant releases per year will have space to grow. Tipuanu Tipu trees produce 10,000 seeds per year, they are wind-borne, very hardy, and can consume atomic nitrogen. We will even carry a few full grown trees down in OPAs to study how well they spread seeds. As soon as we can we will bring in bees to pollinate crops and earthworms to aerate soil. Mosquitos, ticks, moths, cockroaches and fire ants will not be on the guest list.

The mega-green stations will be able to fit one thousand trees each. Since the entire mining economy of mankind will be coming from space by this time we will have a vast demand for food crops in space. Many foods are themselves seeds (most beans, nuts, potatoes) so storing them in space will not be difficult. After the first few years of spreading seeds all over the planet, we will get more precise about our seed containers, even with OPAs being used to bring seeds down accurately to specific, hard to reach locations. We can also use OPA-s to bring Cyanobacteria to every lake or seeds to every island. We could even have an aircraft fly over isolated places and drop containers of seeds or spores at low altitude. If dropped at proper intervals the plants would grow in a grid pattern. We can ensure optimal genetic diversity in each container as well to reduce the effects of plant inbreeding. OPAs used for seeds can be made only with drag parachutes and can be directed down from orbit to save cost. Since they don't have to land intact, they will be simpler to design.

Most plant species on Earth produce thousands of seeds, but they are eaten by animals or land on top of fully grown plants. Due to the complete absence of plants on Mars, wind-borne seeds will have no competitors. If we plant 1 million tipuanu tipu trees in the year after the impact event they will mature by year 10. Each tree will produce 10,000 seeds per year after that. We'll have to see how well the trees adjust to the longer Martian year (we can specifically breed them for this on the megagreen stations). Certain parts of Mars will be scientifically seeded (like farms) but most of the planet will be left to have natural seed growth and distribution.

All of the seeds will be put in very primitive containers in space and pushed into the Martian atmosphere from orbit. The seed containers will be designed to burst open at the correct altitude, spreading their wind-borne seeds over several square kilometers. Aside from picking seeds based on moisture and light, we will use some seed containers to spread life into the ocean as well. Algae, plankton, cyanobacteria, and aquatic plants will be as important to the Martian ecosystems as grass or trees. We can keep small groups of scientists on the surface to study the progress of spreading life in the sea and on land. Eventually we will want animals roaming around Mars but we will give the plants a few years head start on them. I don't want birds eating all of our precious seeds until the planet is green!

After the first four-person crew has scouted everything out, we can fill all of our landing sites with colonists to monitor, control, and study the biologic terraforming of the planet.

We will also include seeding the Martian oceans and large lakes with algae and aquatic plants. This will be hard to model because we won't know the salinity or acidity of oceans before they exist, but we'll conduct what experiments we can before the impact. As the Martian food pyramid is built up we can bring in the next order of animal. In the absence of predators each species will breed to planetary capacity. Then when we introduce a predator for that species it will have plenty of food, and no predators. We will continue doing this until we have brought all of the useful living things from Earth to the Martian land and ocean.

We will create the aquatic version of a greenhouse during phase II by excavating a small pond and then placing a dome over it. We may need to artificially heat it to simulate post impact mars. We can redo these tests after the impact event alters the climate, but this research is better than nothing. We'll continue moving in ice or ammonia asteroids as needed, but the biologic terraforming can get started a long time before the atmosphere is perfect for humans.

PHOTONIC MARTIAN TERRAFORMING

The impact event will alter Mars in a flash (well, a few million flashes.) This will terraform Mars, but it is not the optimal way to keep it terraformed. Mars has no internal convection to help warm the planet or create a magnetic field to push off radiation. It also receives less sunlight than Earth does (or that most Earth-residing organisms expect). As the thermal energy radiates away, enough of it may not be trapped in to keep the environment stable.

Space Corporation will have already built aluminum reflector-1s for moving ice to Venus. Just before these megapili break up near Venus the ARs will be removed and pushed into a gravity assist around Venus and to the Martian Lagrange points. They will be affixed to support beams to make them rigid and they will be angled to reflect more sunlight to Mars.

Even a slight increase in the amount of sunlight on Mars will be a huge help in keeping the planet warm and in boosting the health of plant and animal life settling there. The support beams to keep the reflectors lined up will be the same support beams used to push the ice blade rigs into the Cerean surface. There are already several large asteroids in these positions; we simply need to adjust their rotations with ballistae. Below is a list of asteroids that are already locked in an orbital resonance between Mars and the Sun at the Martian Lagrange 4 and 5 points. There is nothing at the ML1 point, but that object will have relatively low mass. We can build it from pieces of 433 Eros and push it into position.

Table 7. Martian LaGrange objects by size
(Natural asteroids already in place)

ML4 Object	Diameter
1999 UJ7	1 km
ML5 Objects	Diameter
5261 Eureka	3 km
2007 NS2	0.9 km
2001 DH47	0.5 km
2011 SC191	0.6 km

We'll ensure that they only add light to the day side of Mars so that the general intensity of light will be closer to the Earth norm. These points will not be along the line of sight between Mars and the sun, but they will be astride the axis of sunlight so they will be able to supplement the light that the day side of Mars already gets.

There are no objects at the ML1 point, but we need to move an asteroid there to use as the support structure for our large Helmholtz coil. This object will be much easier to make than a large reflexive surface, since the object will be hollow, with only a support structure radiating out from the center (resembling a bicycle tire viewed from the side).

We will place solar panels over the support beams to offset some of the energy that the device will consume. We may need to add nuclear reactors as well; we'll have to see. The same nuclear reactors that are used to construct the coil will remain to power it. The ML1 object will be built after the ML5 and ML4 objects, so we can keep our power generation equipment at ML1 permanently. We will also build solar panels around the periphery of ML1 so that light will be allowed through to Mars, but energy will be available to run the Helmholtz coil. There will be a lot of energy needed to make the ML4 and ML5 objects but almost no power to run them after they are built.

The best design to power the construction of all these LaGrange objects is to place twenty ESR-1s on a non-conducting platform, with a ballista for propulsion, and just fly this ship between different objects as needed. The US military used small generators on the back of trucks or trailers for similar things in Iraq and Afghanistan. By using a large number of small reactors on movable platforms in near zero gravity, maintenance will be very easy. All of the TESRs used for the ice mining operation can have a few more components added to them and be used as standard ESRs to change out on the Helmholtz coil later. With time, we will construct multiple coils on the same object (resembling a slinky in appearance) so that even if something went wrong with coil A, we'd have coil B, physically but not electrically, attached as a backup.

If minor course adjustments are needed over time we can factor this in to the tilt of the coil (just like a sail) or use energy from the solar panels around the coil to keep it in position. Our goal will be to have the Helmholtz coil ready to go and activate it after the atmosphere settles from the impact in year forty. Prior to this, radiation actually helps to heat Mars, so we don't want to restrict it. The main reason we are lowering radiation on Mars is to make it easier for life to spread, and we won't start doing this until year forty anyway.

I should point out that ML1 is not essential for life on Mars, but it is very useful. Our scientists will place sandbags and bricks around their HABs to block most of the radiation and will not have particularly terrible levels when they are traveling outside. If we are going to terraform the planet, we want a place where the ambient level of radiation is close to Earth norm. This is not an Antarctic research station or a military base on a mountain top. We're building a civilization.

FINANCIALS FOR TERRAFORMING

Just taking the figure of $9.6 trillion from thirty years of dominating the Earth's rare metal, gold, and platinum markets, 10 percent of this would be US$960 billion. Terraforming the planets will be primarily done with leftover equipment and technology from the mining operations. By integrating these two activities, we can do both for a fraction of the cost.

The real estate development opportunities are also alluring. Let's assume land on Venus and Mars is valued at one half of Alaska's prices (Alaska is US$2.50 per acre; we'll start Mars at US$1.25). Roughly 50 percent of Mars will be covered by ocean, and 80 percent of Venus. That leaves seventy-seven million square kilometers for Mars and about ninety million square kilometers for Venus. There are 247 acres per square kilometer.

The basic equation I used for this is: (77,000,000,000 + 90,000,000,000) x 247 x 1.25 = US$51.5 billion. Once plots of land are purchased in these places and basic clean water, electricity, and roads are constructed by investors, the value goes to one half of the price in Tennessee (Tennessee is US$1,000 per acre, so terraformed Mars and Venus would be worth US$500 per acre). This now makes the value of both planets US$20 trillion. This will cause everyone to rush in and buy Martian and Venusian real estate when it's still cheap, driving up cost. Even cutting this in half to account for low quality of land and discounts for our grand opening sale, we can still bring in US$10 trillion.

There are many practical reasons why people would want to live on a terraformed Mars. A huge amount of land could be obtained for very little money. That land would be green, pollution free, have crytal clear water on its beaches, stars brighter than any you can see on Earth. A more practical reason for Mars to be settled is the economics of mining asteroids. While a large amount of wealth can be obtained from asteroids near Earth the vast majority of asteroid mass in the solar system is in the main belt. It will be expensive to fly directly from Earth to Mars whenever we want to mine an asteroid. It would be much better to make short trips from Mars to the main belt, then send ore back to earth on unmanned transports.

Some people would live full time on Mars, controlling thousands of space vehicles in Martian orbit. Other workers would fly to the main belt for expeditions to 16 Psyche, 21 Lutetia, etc. We will eventually have small mining stations on valuable asteroids (similar to off shore oil rigs on Earth). These crews will rotate out from Mars for months at a time, sending ore directly to earth with unmanned vehicles. This is common in the gulf of Mexico, or in isolated mining outposts on our current planet. These would fly out as the Martian orbit gets near the orbits of each separate asteroid, then fly back before the orbits get too far apart. We could keep manned colonies on these asteroids at all times and rotate the crews each Martian year. We can also keep autonomous robots there and have humans visit each Martian year for maintenance and upgrades.

There will be ranchers and farmers around the space port to provide food and other consumer products. Very expensive products would be brought from Earth, but economic factors would encourage making more things on Mars. In 1607 American colonists imported everything from England. Today America exports things to the entire world that never existed before Americans created them (power plant components, airplanes, mass produced cars, X-ray machines, MRIs, GPS systems, and computer software). In 400 years perhaps Martian colonists will be selling us faster than light engines and cosmic ray converters. For the 40 years of my plan there is plenty of money to be made by settling Mars and mining the main belt. Mars will need a thriving community of workers and their families to keep the spaceports fed, maintained, powered, and housed.

Many cities on Earth have large populations of sailors. In these cities it is normal for families to remain in the city with their own jobs, or school children attending school, while their family member is away for several months.

Humans could perpetually live on oceangoing ships, but practically speaking few people would want to do that. Mars can be the main hub for mining the asteroid belt. A typical situation would be someone on Earth signing a two year contract to go to Mars. At first mining companies would only offer these to single people, with time entire families would be moved to Mars for two or four years at a time. After the colony is established the entire family of the worker could move to Mars with them. The asteroid miners would spend months away from Mars and then a few months back on Mars. This is how sailors, workers on oil rigs, and many technicians work right now.

Some people will want to permanently live on Mars (particularly famers, miners, construction workers, and people in the tourism industry). Low gravity would also be great for people with joint, back or circulatory problems.

Speaking of which, I'm officially claiming the summit of Olympus Mons. I need space for my five-star restaurant (Top of the Worlds). I'll make this official when the Space Corporation Charter is approved by the US Senate, but I wanted to let everyone know up front.

Anyway, with making US$800 billion a year for all thirty-two years, and US$10 trillion when the planets are done, this is US$45.6 trillion. We can secure our start-up costs for probably US$10 billion in cash (assuming the technology outlined in my plan will be exchanged for stock shares). I'd also like to add that this plan was designed around US government owned equipment and technology. It is very likely that the US government would own the majority of the stock, since stock will be used to compensate the US government for its resources. It's reasonable that 75% of the stock would be owned by the US government, with the other 25% going to private investors who buy shares. Assuming that %30 of the money covered operational costs, and 10% went to long term terraforming operations, we would have 60% left for our shareholders. This would be $20.4 trillion dollars. That could pay off a huge part of the national debt.

The whole point of stock companies is buying a share of future profits. While most cities around the world are named for tribes, natural phenomenon, or geography, American cities are frequently named after the wealthy businessmen that created them.

There is a Ford City, a Carnegie Pennsylvania, (Pennsylvania itself is named after Thomas Penn). Countless local towns, counties, even state capitals are named after prominent businessmen who founded the cities, prominent politicians who assisted in founding a city, or a name that would impress businesses men or politicians.

I don't know which communities, companies, or private citizens would be interested in buying stock in Mars Corporation or Venus Corporation. Having said that, here are some names that are so good someone really needs to invest in Space Corporation to ensure they are used: Branson City, Theilville, Musk City, Perrotville, Cubanville (named for the billionaire, not the poor socialist dictatorship), New Houston, New Atlanta, New Milan, New Nagoya, New Frankfurt, New Bangalore, New Bristol, New Moscow, and Trump Land.

Despite these estimates, I will admit that it is a possibility that even if the US government passed the corporation charter, private investors may not be willing to fund this endeavor. The core problem with funding outer space projects so far is that profiting from it has been illegal, so investing in it has been pointless. Even if none of the investors on Earth wanted to touch this with a ten-light-year pole, it is a morally depraved idea that *trying it* should be illegal.

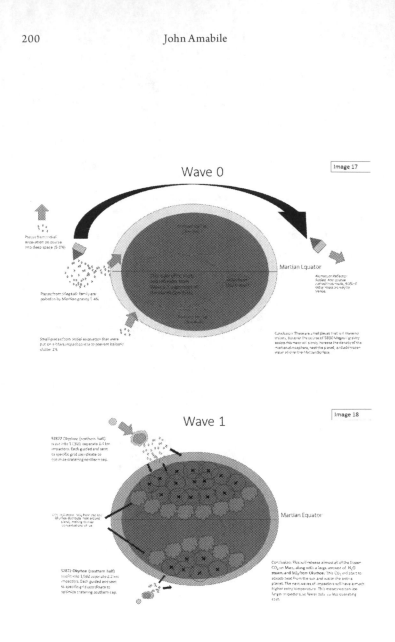

2060 Chiron (200 km diameter), 174 P Echleklus (84 km diameter), and 8405 Asbolus (84 km diameter) are broken up 7,000 km away from Mars. Chiron is cut into 8 pieces (named Chiron 1-8). Each piece will have blast tunnels added and come in one day apart. When they are 7,000 km away from Mars a fusion device will break each piece into thousands of 600-800 m pieces . Inertia and Martian gravity will pull them all into the Martian atmosphere where they will melt on the way down. No precise grids planned for each object. The density of the air will go up with the 5E18 kg of mass from these objects.

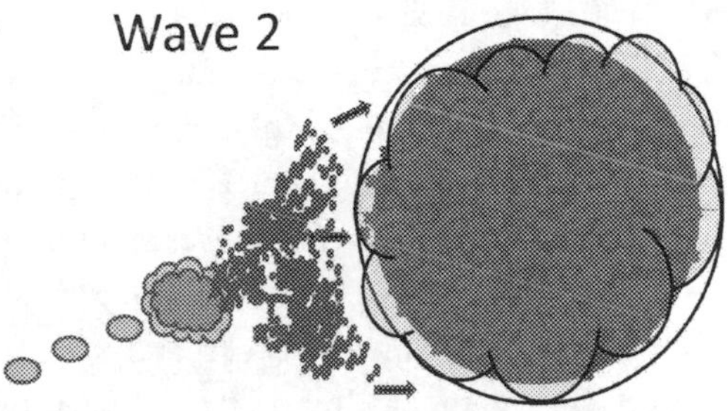

Wave 2

Image 19

Conclusion: The atmosphere has all of the nitrogen it needs, it is super heated, and it has an ocean worth of vaporized steam.

One day after Wave 2 ends the same method is employed on Jovian moon Amalthea and the asteroid 596 Scheila (both primarily water-ice). Entry temperatures will be above 6000 C, so water will disassociate into Hydrogen and Oxygen. This will produce 2E18 kg of oxygen gas and 2 E17 kg of Hydrogen gas (1 E18kg gets absorbed in ocean and rocks, the rest is in the atmosphere).

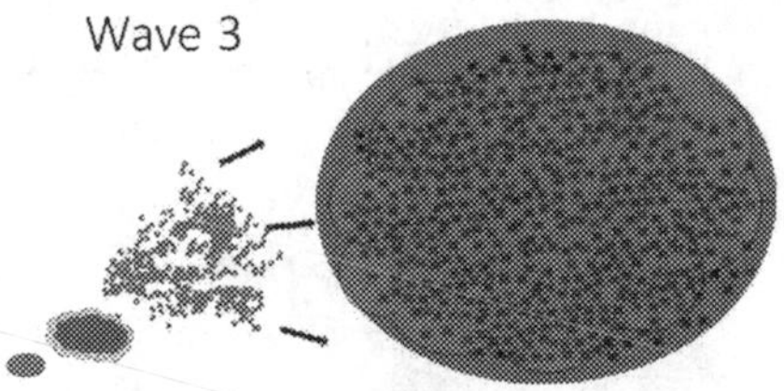

Wave 3

Image 20

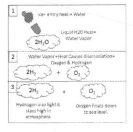

Conclusion: Martian atmosphere has 4E18 kg of nitrogen gas and 1E 18 kg Oxygen gas. This is same mass as Earth, gravity and volume will cancel out so air pressure will be earth norm. CO_2 and Methane will be higher than Earth making the planet the same temperature with less sunlight.

TERRAFORMING MARS CONCLUSION

If half of Mars were covered in water, it would still be twenty times the landmass of the continental United States. Far from a fantastical government project that costs $100 billion to plant a flag, I would like to make trillions of dollars while planting civilizations. Almost all of the technology used in my plan is forty years old, and a lot of it is being guarded in expensive Department of Defense warehouses right now.

We know from Earth's history that transportation technology comes from a market that has a lot of things to move. The transcontinental railroad across the United States was only possible because the railroad companies were able to sell all of the land along the tracks. They helped create farms, towns, cities, and factories. With products and people to move, both the geographic area and the railroad thrived. Since we live in Solar System 1 planets and asteroids here will be purchased first and for the highest price. The other 800 billion solar systems in the Milky Way Galaxy can be sold to "Light+ Corporation" for a nickel. The shareholders of Light+ will then have 800 billion reasons to perfect faster than light propulsion. Spending 25 billion dollars to reach the 200 known exoplanets makes a lot more sense if you've won the bids for those 200 exoplanets on eterrafrom.com.

Without ownership rights of the Milky Way Galaxy the only economic use for faster than light space propulsion would be bored tourists flying donuts between Earth and Pluto.

CHAPTER 9

Major Political Points of Contention

The previous chapters of my book were concerned mostly with the clear issues of physics, philosophy, chemistry, metallurgy, and economics. Now I'm going to delve into the complicated stuff: politics.

GLOBAL WARMING

Global warming attracts an enormous amount of attention from the press, the movie industry, musicians, schoolchildren, and politicians. I personally think the extent of anthropogenic climate change to be exaggerated and the media coverage of it to be wildly partisan. I am also stunned at the illogical and politically charged solutions that are usually recommended by environmentalists. Just like space colonization, global warming has fallen afoul of the usual government approach to solving problems: charge taxpayers a fortune and hope they don't notice you didn't solve the problem.

Most carbon-producing activity on Earth is being pushed to China, where there are no elections and no environmentalists. Many environmentalists may not realize this, but China is on our same planet, is the largest CO_2 producer on this planet, and emerges completely unscathed (rarely mentioned) from public discourse on climate change. Global CO_2 production has gone up consistently for two and a half centuries. In China it is going up very fast, in the west it is going up slowly. I logically deduced China's CO_2 production from their economic growth rate, and looked it up to confirm my suspicion. I never learned about it in school or saw it on the news. As free societies lower CO_2 production (and jobs), we just move CO_2 production (and jobs) to unfree societies. Even assuming that 100 percent of the global warming arguments made by western environmentalists are correct, moving CO_2 to a different part of the same globe will only heat the globe more evenly. This is how you cook a good omelet, not how you save a planet.

One of the reasons that the United States borrows money from China is because China has no environmentalists, no environmental laws, no freedom of speech to discuss environmental issues, no elections to pick politburo members concerned about the environment and no desire to lower CO_2 production. This helps China manufacture a lot of things cheaper than more environmentally conscious countries. They also have more money for their defense budget (or offence budget if you live in Tibet, or the South China Sea). They are still heating the globe. We aren't sacrificing millions of jobs and trillions of dollars to cool the globe. We're sacrificing millions of jobs and trillions of dollars so the Chinese get to warm it instead.

Even the terms *global warming* and *global climate change* are misleading. The issue has only been pursued as "congressional district warming." China alone accounts for 97 percent of the rare earth metal smelting capability on Earth. Vast amounts of manufactured goods (which are what the mines are producing material for) also come out of China. This doesn't make the state-controlled press, state-controlled schools, or state-influenced movie industry in China. In the West, we have no excuse. The claim that China is larger so it's okay that we don't discuss their CO_2 production may be politically correct, but it is clearly not scientifically correct. Even if all factories in the west closed and all western citizens became subsistence farmers, the planet may cool for a few years, but Chinese CO_2 production would continue to go up. Granting all Global warming arguments to be true, why is protecting the lives of human beings on Earth less important than cooking the planet evenly?

Ultimately, CO_2 is generated by human beings and human economic activity. When relatively clean coal in Kentucky is restricted, Americans pay for much dirtier coal to be shipped from Indonesia (which I've seen with my own eyes in a power plant in New Jersey). When Australia subsidizes ranchers for not cutting down trees, the world price of lumber goes slightly up, and an extra acre of the Brazilian rain forest is cut down. Since we share the same planet with Brazil, Indonesia, and China, "congressional warming" concerns just move the CO_2 production around the globe—they don't slow it. Developing nations have much less efficient technology than developed ones, so they generate more CO_2 per unit of energy. This cooks the global omelet better, but it's not saving anything or anyone. No one should have any illusions that humankind has no effective global government. Baring large-scale warfare, we never will.

One of the unique attributes of my plan is that it would create a free-market alternative to mining rare earth metals *on Earth*. We really need to call them rare China metals. They are not owned by an Earth-government that does not exist. Large parts of the smelting and mining industries on Earth would be shifted outside of Earth's atmosphere.

Using 2010 numbers, there were 2,500 metric tons of gold mined on Earth. To smelt one kilogram of gold, 11.5 tons of CO_2 are produced. If we set this number to the 120,000 tons of rare earth metals produced on Earth, we get the figure of 1.38 billion tons of CO_2 produced per year. Rare earth metals are probably much more CO_2 than this (the smaller the concentration of an element in ore the more energy is needed to melt 1 kg of that material out). Even taking the 1.38 billion ton number this is equal to all of the CO_2 produced by Australia, France, and Poland combined. Or should we stay cooped up on Earth and see how energy-efficient lightbulbs work out?

Were space mining to become a viable market phenomenon, we could mine gold on Eros 433 (zero carbon footprint) and then drop it directly to its destination in the United States via an OPA-1 (zero carbon footprint). Democratic states with large mining sectors like South Africa will not be hindered economically by this, as their status as a democratic state and a huge source of mining expertise will allow them to move in on the share of the mining market currently controlled by China. The influx of gold will lower its price, and the companies that deal with gold as a finished product will benefit (since their operating costs will go down). We will start with the most expensive metals, but eventually all but the cheapest metals will be mined, smelted, and transported outside of Earth's atmosphere. Someday the idea that we conducted mining on Earth will be as strange as the people who once grazed cattle on Manhattan Island.

Aside from global warming, there are a lot of secondary pollution effects from mining here on Earth. Trace amounts of heavy metals from mined ore can run with rainwater, leading to heavy metals and sulfuric acid contaminating soil and drinking water. Also huge amounts of water are used to filter metal for mining, and much of this water becomes polluted in the process.

Since my method of acquiring metal is cheaper, it will be adopted on a voluntary basis and only by each person, corporation, or government in their capacity as a consumer or producer. Besides the Space Corporation Charter, we won't need treaties, speeches, concerts, or movies, none of which work in unfree societies. All we'll need is people who desire to make money and are willing to buy cheaper space-mined metals. I don't think it will be hard to find a few billion people like this.

There are also many elements on Earth (like neodymium) that are useful for alternate energy production but prohibitively expensive (US$2,500 per kilogram). By mining asteroids in space, we will lower the cost of many elements that consumers on Earth use. Green technology will be cheaper as specialty metals become more abundant. This also applies to super-conducting material that will make power transmission on Earth more efficient.

After the climate and hydrosphere of Venus and Mars are rendered earthlike, covering these worlds with plants and algae will be essential in making them marketable for Space Corporation's customers. We will be able to spread life fastest in the sea, with cyanobacteria, phytoplankton, baitfish, larger fish, and then fish-eating mammals like dolphins. We will exclude the poisonous Australian jellyfish, because they prevented me from swimming during my vacation in Cairns, Australia, in January 2011.

Once enough plants have grown on the surface, we will land thousands of fast-maturing marketable herbivores (pigs, cows, buffalo, water buffalo, yaks, turkeys, horses, camels, etc.). We can have entire islands reserved for Earth-endangered species or simply spread them all over one of our planets.

On Earth, there is resistance to environmental measures because humans have an adversarial relationship with many animals. Conservation efforts for many great apes are hampered by the poverty of Indonesia and the political instability of Zaire. We can tailor design islands or continents on new worlds for these species and transport small breeding populations there.

If we get VASIMIR propulsion or NSWTR working, we can even sedate young whales, launch them into space, and then rush them to another world in a week. After entering the atmosphere, their capsule will deploy a parachute, and they will float down to the ocean (OPA-76?). We can have most of the container break off after atmospheric entry, leaving a thin cover that is soluble in the Martian ocean. The minimum breeding population would be around thirty, so this would not be easy or cheap to do. It will be the best thing to happen to whales since Rockefeller made whale blubber unprofitable. Some environmentalists claim that they would like a world without people. My goal is that one day we'll be able to sell them two.

Finally, for the radical doom-and-gloom types that are expecting some wild, apocalyptic, Hollywood climate-change event, what better solution could there be to global warming than for mankind to acquire more globes? In one hundred years, when someone on a microphone is screaming about the end of the world, hecklers will get to shout, "Which one?"

WASTE DISPOSAL IN SPACE

Another wonderful innovation of Space Corporation is waste disposal. Imagine how simple it would be to get rid of garbage if the sun were at the bottom of every trashcan. When you wanted to dispose of something hazardous, the sun's heat and gravity would take care of it for you—no pollution, no risk to drinking water, no bacteria, nothing. On Earth, having the sun in a trashcan would only occur in a hallucination, a fantasy novel, or a politician's campaign promise. In space, on the other hand, as long as you design your ship's trash compactor adjacent to your airlock, it will simply be one button away. Any ship on a sunward trajectory (within a few degrees of the sun) could dump trash into space, and that trash would eventually float into the sun. The only caveats would be to ensure that no planets, asteroids, or children on a space-field-trip are in the ballistic path you are using.

Speaking as a US Army veteran, it is very expensive to hit something in space on purpose. A few flight paths reserved for sunward trash would be more than adequate. (Earth is moving at twenty-four kilometers per second.)

We can also build a waste-recycling station in Mars orbit. Once nuclear fuel decays to the point where it is not useful for industry, we can store it in a space station, where what's left of the energy will induce an electric current and help produce hydrogen propellant. Even if an accident occurred, it would be forty million miles from Earth and would result in a slightly radioactive thing crashing down on a mostly uninhabited world. Remote-controlled robots will handle all of the actual atomic fuel, so there will be no health risks to scare people into disposing of useful resources. It won't matter if the rate of power production is low; every bit of energy will help produce more fuel.

Due to the high cost of exporting anything to space, we will be big on recycling. But if something has to be thrown away, we're never more than one button from an infinite trash can.

ORGANIZING PRIVATE PROPERTY CLAIMS IN THE SOLAR SYSTEM

The planets will be established as subsidiary corporations: the Venus Company and the Mars Company. Perhaps someday there will be a Mercury, Saturn, and Ganymede company, but I'm pacing myself. Anyway, having them as different subsidiaries with fixed budgets would prevent conflicts of interest. We want to sell plots of lands on these worlds, but we must be careful selling land until after the appropriate phase of terraforming. I don't want anyone's plot of land to be covered by a lake when they arrive on Mars. If that happens, they'll lower our merchant rating on eterraform.com. After kinetic terraforming is complete, we can sell land claims and charge extra for custom terraforming.

We will meet with our clients and make a plan for biologic terraforming of their property. All of the land on Mars and Venus will need to be seeded with plants. It would be fine if some of our customers wanted us to seed their future holdings with specific species. If a Japanese company buys Beta Regio and wants mackerel and a specific species of seaweed planted off the coast, we can certainly oblige them. If an Argentine company buys part of Mars south of Tharsis Mons and wants to set up a cattle ranch, we can plant alfalfa for them a few years in advance.

Space Corporation will retain a small amount of land along future roads, seaports, spaceports, airports, canals, bridges, and any other pieces of planned infrastructure. Just as on Earth, we want these plots of land to all have access to a public road and a water source.

When we're done, we can divide Mars into one million plots of land, with roughly equal value. Coastal regions and regions with known valuable resources will be smaller, where relatively barren or inland regions will be much wider per plot. We can auction all of these plots off and then split the money among the investors. We can let some investors buy preferred stock in the planetary corporations so they will get first pick on land claims prior to the public offering. Like any investment, the people that buy this stock first can get it for less money. Since this is my idea, I obviously get preferred stock (summit of Olympus Mons!).

While citizens of democratic states would insist on any government organization publishing all of its survey data, private businesses regard information as property. Some percentage will be released to the public to create general interest, but detailed information would be given only to prospective buyers. The company may decide to disclose everything; the information will belong to the corporation to use in whatever way they think profits them.

SPACE CORPORATION AS A LEGAL ENTITY

Throughout this work, I refer to Space Corporation in the singular, but this could certainly be organized into a multitude of separate companies (one for mining a specific asteroid, one for terraforming Venus, etc.). There would have to be a small central business planning and coordinating the terraforming activity, as well as handling the terraforming budget. Any large global business delegates most of its day-to-day decisions to its regional offices, and the highest level only coordinates between the smaller sections, so this will not be very difficult.

Separate entities will have to pay their 10 percent of profit to Space Corporation. Space Corporation will use this money to pay other companies to terraform Mars and Venus, as well as for capital costs for space infrastructure (green stations and propellant stations).

Hopefully, short term profit operations can be nested with long term terraforming operations. If Company A makes $30 billion mining neodymium on asteroid 21 Lutetia, they will have to pay Space Corporation $3 billion. If 21 Lutetia also has a good aluminum deposit, Company A could make the 3,500 aluminum reflectors using their same smelting facility in orbit around 21 Lutetia.

If we took bids from all the other companies for doing this it could be much more money. Company A already has the position and the infrastructure in place. Company A may be able to do this for $1.5 billion but negotiate to have it counted as $3 billion for their Space Corporation fee. This will all be voluntary. If a company just wants to make money and have nothing to do with terraforming or infrastructure, they pay their percentage, and that's it.

Space Corporation shares will be publicly traded on the Space Exchange, though only citizens from free societies will be able to buy anything on the space exchange. As companies are certified as being from free societies, they can place their shares on the Space Exchange as well. Most space businesses today are American, British, or Russian, and all of them can be admitted to the Space Exchange if they wish.

Whether the companies want to merge and all be on the same stock ticker or be separate and accept the 10 percent terraforming charge themselves will be up to them. Working together will be more profitable for all of them, and the board of directors will ensure compliance with the company charter.

All the property in the solar system, except for the sun, Earth, and anything man-made, will belong to Space Corporation. Space Corporation will sell or lease this property to other entities to help generate income. At first, Space Corporation will be pressed for money, and the prices of asteroids, planets, and moons will be very low. Any start-up business has a lower stock price than after that business has succeeded.

The Space Exchange will allow a legal entity through which property, labor, and money can flow. If a company wants to mine ice on the moon, they can buy the moon, lease mining rights, or offer their services and solicit funding through the Space Exchange. This is how economies work on Earth (on those occasions when they do work). This is the only way it can work in space.

Since there are already several companies exploring space mining and space tourism, these companies would benefit from Space Corporation and the Space Exchange, because they are in the front of the line with expertise and technology.

THE RATE OF EXCHANGE FOR RESOURCES VERSUS STOCK

The rate at which resources can be exchanged to Space Corporation for stock will of course be a complex issue. The weapons were made at a great expense, but there is no free market value for them. After World War 2 a lot of military equipment was simply auctioned off, or sold to other nations. Most of the technology I've highlighted in this plan is only useful in space (intercontinental ballistic missiles, plutonium bomb pits, and space debris). By letting companies in the space exchange bid on these items (with stock shares) we can ensure the best value is obtained for them.

A big problem with arms reduction is that America and Russia have to reduce arms simultaneously, since neither wants to unilaterally pull 2 percent of their weapons offline. By selling disassembled weapon components to Space Corporation in exchange for stock, both nations will have a face-saving way of reducing their armaments and save money on their maintenance and security. Any democratic state on Earth will be invited to contribute resources, but most are owned by Russia and America.

Frankly, Russia would gain much more from contributing some of its nuclear weapons than America, since it has more systems and heavier rockets. There is very little rational military use for possessing such a large number of atomic weapons. Certainly, as the systems age, there's no reason not to remove the warheads and sell the missiles to Space Corporation.

Were disassembled atomic weapons between the United States and Russia used for demolitions or as fuel, there would be zero impact on the military balance of power on Earth, other than making the United States and Russia trillions of dollars and best friends. I am assuming it will be politically complex to do, but many people would like to be rid of these devices, if even the ones in storage. Of course, some provision will be added to address equipment that doesn't work (they'd have to give us a refund.)

Space Corporation will purchase space debris from whomever wishes to sell both for cleanup and for recycling as components in future space missions. Of course, all space debris will be purchased in the space exchange before it is gathered and melted down by our proto-ballista. The exact rate will be adjusted by distribution, with one massive piece being worth a high rate and a large disbursed cloud of debris being worth a lower rate.

By current international law, nations own their debris in orbit, so they will have to be compensated or sign their debris over freely. They are also liable for damages in the event of a collision, which is only a matter of time. It would cost US$7.7 billion to fly up the same amount of mass that is circling our planet already.

We'll have the delicate issue of citizens trying to cross state borders to work for a private business or even a different government agency supporting the Space Exchange. If the chief

Russian space-walking expert were to move to Florida and wanted to work directly for Space Corporation, Russia would reasonably protest. Each nation's space administration should be allowed to claim part of the wages if highly trained personnel wish to cross national borders without their government's consent (perhaps 10 percent). Obviously, I support the right of states and people to voluntary relocate as long as both parties consent.

During the research for my book, I was surprised to learn that all three of the scientists that developed the first nuclear fusion bomb were immigrants who fled communist-controlled Eastern Europe. Dr. Van Neuman and Dr. Teller were both born in Hungary, and Dr. Ulam was born in what is now Ukraine. Suffice to say, these men were smart enough to move to a free society before they started complicated developments in physics. Similarly, of the four men most responsible for America creating the atomic bomb, two of them were actually from nations the United States was at war with at the time: Albert Einstein was born in Germany, and Enrico Fermi was born in Italy. Enrico Fermi actually snuck out of his Nobel Prize ceremony in Sweden to come to America (he was forbidden to leave by the Fascist government).

Not only do I want people to be free to move to space, but I want free people to be the ones moving to space.

ANTI-HUMANITY ARGUMENT

The argument that mankind is an immoral life-form or that it is not ethical for us to alter our environment I hold to be absurd. Except for a tiny number of hunter-gatherers, all human beings on this planet alter their environments to improve their lives.

The group of people in America who are best at not altering their environment, besides the dead, are the homeless. Far from either political party holding the homeless as a moral example, we are usually taught to feel shame that we haven't altered their environment more. People have been building things on this planet for tens of thousands of years. People like having food, security, technology, access to information, and transport. I do not know when it became acceptable to protest means while demanding ends.

Virtually no one in our society voluntarily lives in a state of nature if they can avoid it. Everyone in America is free to live in the woods and not use electricity but virtually no one wants to do that. We actually have laws that guarantee access to things made by altering our environment (food from modern farms, clean water from modern treatment plants and electricity from power plants).

My grandparents grew up in a village that had not changed for a thousand years. No capitalist decided to burden them with running water, electricity, sewers, cars, trains, banks, telephones, machine-produced clothes, education past the third grade (in their obscure Italian dialect), medical care, policemen, or peace. They were not ecstatic to be free of material possessions (especially food, doctors, and sewers). It was miserable, and they left as soon as they could.

No one will be forced to leave Earth, and no one will be forced to purchase products retrieved from space. Everyone will benefit from buying cheap products from space, and that's almost the number of people that will want to do it. If someone tells you that mankind is evil, unless they are a rude robot that means they are evil. If the first point a person brings up is that they are evil, why would you care what their second point was?

SCOPE OF GOVERNMENT REGULATION

The use of asteroids as impactors and of atomic/nuclear devices for mining will be legal but strictly regulated. To be clear, I would like to say that even typing the word *regulation* is painful to my fingers—it's like casting ten tiny chains made out of ink. But I do think there is a tiny sphere for just actions of government, and in space, asteroid collisions and atomic mining will be a few of the activities that warrant regulation.

Members of the military from the state contributing the fusion devices will personally escort the devices into space and will be the only people permitted access to them. This will be essential for terraforming Venus and Mars. This will be handled just as civilian companies must get blasting permits on Earth for construction work or to handle hazardous materials.

Both the US and Russian militaries spend a lot of money tracking the thousands of objects in low Earth orbit. We created this ability because the nuclear war we were prepared for would have largely been fought in space. During the Cold War, we thought that an enemy atomic bomb would be hiding in those debris clouds, so developing the radar technology for tracking them was not an idle exercise. As space debris is cleaned up and the divide with Russia only comes up during athletic competitions and James Bond reruns, these installations can serve as an impartial tracking system to adjudicate disputes between shipping companies, mining companies, and tourists. Space Corporation will be responsible for tracking the positions of the objects in the solar system in the event that legal action comes from collisions or course adjustments.

Sunlight will be considered the property of the object that would naturally receive that amount of sunlight. It will be illegal for someone to build a sunshield and hold the light supply of Phobos hostage. This will be treated the same as landing on

Phobos and physically stealing someone's property. However, there will be a few exceptions to this rule:

1) Sunlight can be diverted for the sake of terraforming, but approval from Space Corporation itself will be necessary to do this. (ML4 and ML5 will be used to terraform Mars).

2) An incidental loss of sunlight (a ship or small asteroid passing along the line of sight with the sun) cannot be challenged in court. We'll set a specific time and surface area limit for this.

All of the Martian LaGrange objects will be owned and operated by Mars Corporation, as a subsidiary of Space Corporation; Venus Corporation will own the Venusian LaGrange 1 object. This will be similar to the way power companies are organized in the United States, as private companies regulated by the state. Except in space these will be local companies regulated by Space Corporation. The goal will be to design these objects so that they have virtually zero budgets or human interaction. This will spare them from political infighting or mismanagement. We are thinking more like Roman aqueducts powered by gravity and less like a bureaucracy powered by taxation.

While natural property in space will be very cheap, some asteroids will be more valuable than others (especially after the data from the reconnaissance phase comes back). Space Corporation will want short-term capital. It will sell, lease, or rent these asteroids to anyone in the Space Exchange. As the space infrastructure develops more and more, the value of these asteroids will go up, and more entities will form in order to purchase them.

While I have specifically addressed Venus and Mars in this work, I want all property in the solar system eventually sold on the free market. If someone wants to get a team of investors together and buy Mercury, they should be free to do so. I have no idea why anyone would want Mercury, but I don't assume my intelligence so superior to mankind's that I would impose my views on anyone.

The people in Kitty Hawk probably thought the Wright brothers were a bit old to be making a large toy bird on a beach. Fortunately they lived in an age before entrepreneurs and governments were enemies.

We will create formal rules for accidents in space (similar to laws for cars or aircraft on Earth). Anyone will be obligated to rescue distressed crew in space. Fair compensation for this will be imposed on the receiving entity, since it could cost millions of dollars to divert a ship on a mining expedition. Hopefully existing space agencies can conduct this service in space as a kind of 'Space Coast Guard.' We can also position simple unmanned ships with rescue equipment and emergency rations near populated areas or common flight paths. Basic safety equipment will also be tied into the Space Corporation Charter, but only for interactions between parties, not for the safety equipment a company wants on their own machines. Decrepit equipment that poses a safety risk may be banned by the board of directors (which occurs in rare cases on Earth with mechanical defects). Radio frequencies and docking ports must be standardized for rescue operations, but this will also facilitate commerce and should illicit little opposition.

Speaking of legal disputes, I'm sure there will eventually be low crime rates in space. I've no doubt that two men in Musk City, on the northern shore of the Hellas Sea, will consume a few beverages that contain alcohol and proceed to get in a fight over a woman, who doesn't like either of them. Large companies and small cities arrange their own security; it will be easier to do with small numbers of well vetted people in space.

None of the signatory powers to the Space Corporation charter may pass laws restricting their own citizens' liberty to buy property in space, and no nation can confiscate space real estate. Imagine the following scenario: A group of colonists from Cortina sets up a successful mining business on the Martian moon Phobos. The government back on Cortina passes a law that raises taxes on them or even nationalizes their property. Then all the colonists on Phobos put on Klingon Halloween costumes and dump all of the imported tea (or whatever they'll be drinking, probably Tang) into space. They also adapt their ore catapults to fire at approaching ships. Crazier things have happened.

Investment in space will be enormous if space commerce is taxed at 10 percent but Earth business is taxed at 35 percent or higher. This will push all practical economic activity into space. This was a core idea behind the growth of the United States (with a 0 percent income tax rate before 1916) and was the engine that drove investment and immigration from Europe starting in the seventeenth century.

Misconduct by signatory powers or companies will result in fines, restrictions, or banishment from the Space Exchange, with the stockholders of the company as guarantors.

Of course if companies go into space without Space Corporation, we will not use force to stop them, but we will not allow them to use our technology, infrastructure, or expertise. We'll make so much money doing this that no one will be able to compete with us.

LIMITING SPACE CORPORATION TO DEMOCRATIC STATES

The big difference between the colonization of space and the colonization of Earth is that most of the world was colonized by autocratic states and in an autocratic way. The Spanish Inquisition was torturing and oppressing ethnically European, Catholic, Spaniards. How surprised can we be at how the conquistadors treated the Aztecs? Governments who use force on their own citizens find war against foreigners a wonderful distraction.

After ten years of democratic elections, a state can be admitted into the Space Exchange. This will peacefully encourage democracy. The fall of the Soviet Union in 1989 created dozens of democratic states. Perhaps in forty years we can double that.

John F. Kennedy once said, "Those who make peaceful change impossible make violent change inevitable." I don't want a war or revolution on Earth affecting the company's stock. Any joint stock company is a gamble. Free countries are a safer bet than normal countries.

RESOURCE-USE ARGUMENT

While researching for this book, I was shocked to see posts on the Internet where a child asked a basic science question about terraforming and a college professor replied that mankind should learn to live with our resources on Earth before we go out to space. If someone sends me mankind's mailing address, I'd be happy to write them an angry letter about living within their means.

If mankind had to unanimously decide on your breakfast you would starve to death before 1% of them heard about it.

"Mankind" is actually seven billion separate human beings, each with his or her own independent mind, language, interests, enemies and friends. Mankind (unanimously) has not done a single thing in all of human history, and it never will.

Even unanimous UN declarations don't reflect mankind, since many states in the United Nations have no free elections and all known unanimous elections were not free. It would be impossible to get seven billion people to blink at the same time, and that's something they all agree on and do naturally.

When it comes to highly divisive issues like property rights, ownership of the means of production, ownership of consumer goods, and wealth redistribution, it is rare that a married couple has the same opinions, let alone the equivalent of 3.5 billion married couples (simultaneously, over the same property, across two hundred national governments, thousands of ethnicities, professions, religions, and hundreds of thousands of local and regional governments). I could mention the dozens of national or regional governments openly at war with each other, terrorist groups, organized crime groups, and the unorganized criminals in the world. Hopefully you appreciate that assuming all of mankind will agree to share a pronoun is ridiculous.

Business, on the other hand, empowers individual people to do whatever they wish without harming others; it is an actual, tangible group. I could easily figure out how many people drink Coca Cola, or wear Nike, or drive Fords etc. No businessman has ever waited for mankind to approve his actions. He only needs enough customers to get started.

It is impossible for mankind to do anything unanimously, but business can do virtually anything with a handful of consenting people. Amazon, Verizon, Apple, UPS, and Microsoft, have made it possible for you to buy millions of items across huge parts of Earth and have them shipped to you in a few days. No war, protest, rebellion, or election is required for you to buy something on your iphone, and have it flown halfway across the world to your door.

This is the beauty of a free market system—that it creates a nonviolent way for people with different backgrounds, views, and ideas to exchange property, labor, technology and information to their mutual benefit. A majority vote of the US Senate and a few investors would be the minimum to open up space for everyone on Earth. If the other 6.7 billion people on Earth think it's a bad idea, they can just pay one hundred times more for the rare China metals. It's wonderful that so many people are free to buy or sell their products. It's terrible that it's illegal to do that in 800 billion solar systems in the Milky Way Galaxy.

Many of the current mining nations of Earth will have an economic transition if space mining were to go into effect. America, South Africa, Australia, Canada, and Russia will be free to join Space Corporation, where mining expertise will be in high demand. The smelters on Earth that process the ore into consumer goods will get even more use as the quantity of these metals increases. China currently has a near monopoly on rare earth metals (they are, after all, only rare *earth* metals).

As long as transactions are consensual, the unanimous consent of mankind is not necessary. All we need for this plan to succeed is for the US government to rescind the Outer Space Treaty, to transfer old equipment to Space Corporation in exchange for stock, and for a few thousand private investors to buy stock. Far from requiring mankind's consent, we wouldn't even have to tell most people on Earth.

Normal media coverage and commercials will keep people in free countries aware of our day to day operations. Occasionally we can put up public notices reminding transitioning nations that free elections will allow them to join the space corporation.

In year 20 the megapili going to Venus will pass Mars and will be visible from Earth. Just as aircraft produce clouds as contrails without anyone being shocked someday the night sky will be full of 5500 stars (really megapili) en route to Venus. Even OPAs coming in by the hundreds or thousands will make shooting stars a common site.

People in unfree societies will have little or false information about what space corporation is doing. But not an ocean of tears or all the guns in the world will be enough to stop billions of people from looking up to free space. We'll plan a short public service announcement at this time so that every resident of a dictatorship will know what is happening in the heavens. Nothing preachy or philosophic, just a brief note, in a thousand languages, across six continents, so that no one is shocked by the changing sky.

The Americans are moving the stars

Conclusion

America was colonized with a little protection from the British navy and a huge number of private businesses. For almost the entire period from 1607 to 1916, there was no direct taxation in the United States. There were a few months of indirect taxes in 1775, but this was unpopular, and the policy soon changed the government. This is precisely how I suggest that space be colonized, except that we can skip all the nasty wars, because there are no people to fight in space. The only way dictators can leave Earth is if we carry them up, feed, and propel them. They'll have a better chance going to space by circus cannon.

The United Kingdom had a hereditary monarchy in 1776 and a hereditary upper house of Parliament. Thomas Paine once joked that having a hereditary government is as smart as having a hereditary mathematician. It is only from Britain's lower house, the House of Commons, that all modern democratic forms of government on Earth find their origins. Thomas Paine praised the partially free parts of the British constitution; the American Revolution was to free our House of Commons from Britain's House of Lords.

Only monarchies participated in colonizing the New World, and the difference between the limited monarchy of the United Kingdom at the time and the unlimited monarchies of France and Spain is the source of most material differences between the United States and the rest of this world. Imagine what all the democracies of the Earth could bring forth?

The introduction of space mining will see a drastic decrease in the price of rare *earth* metals. This will make superconductors, engines, computers, and transistors—all current technological items—cheaper and more affordable for mankind. Less energy will be needed to refine and smelt these items, and more of that energy can be expended in space with nuclear fuel instead of on Earth with fossil fuel. While there is a big debate on when Earth's resources will be exhausted, conflicts will always break out over resources. Whoever gets to space first gets the cheapest price. How's that for building a consensus?

The vast majority of wars in human history were an unfree country fighting another unfree country. Imagine if bronze and the wheel were introduced to South America by Starbucks instead of conquistadors sent by absolute monarchs? Almost all of the economic growth, all of the inventions, roads, farms, factories, and technology that has turned mankind into an atomically powered, space-faring society was the work of the free market.

Other than a handful of good accidents from WWII weapons (like microwaves and satellites), all technological innovations of the modern world have been created by businessmen in expectation of profit. The printing press, steam power, railroads, electricity, airplanes, petroleum, the moving assembly line—these were not gifts from benevolent governments; these were items created by self-interested businessmen. Gutenberg had to take out a loan from a bank to create the printing press. Would he have done that if printed words were the collective property of mankind? Should we be surprised that we are moving to space as fast as condo construction on Aphrodite Terra?

Who knows what riches and achievements will come from a universe of free people? I've spent a lot of time writing this book, and I can scarcely imagine it. The term *Old World* will get an entirely new meaning.

With no commitment of money, with no real loss in military technology, and with literally the entire universe to gain, what honest reason is there to not open up space for man?

In 1803, Thomas Jefferson said of America, "We will take Old Europe by the hand and walk beside her. What a colossus we shall be." It bothers me greatly that the ability to think of the future seems to have passed with our ancestors.

Printed in the United States
By Bookmasters